WETLANDS OF RHODE ISLAND

by

Ralph W. Tiner
U.S. Fish and Wildlife Service
Region 5
Fish and Wildlife Enhancement
One Gateway Center
Newton Corner, MA 02158

SEPTEMBER 1989

Published with support
from the U.S. Environmental Protection Agency
Region I, John F. Kennedy Federal Building, Boston, MA

This report should be cited as follows:

Tiner, R. W. 1989. Wetlands of Rhode Island. U.S. Fish and Wildlife Service, National Wetlands Inventory, Newton Corner, MA. 71 pp. + Appendix.

Credits:

Credit is given to the following sources for permission to copy some of the illustrations found in this book:

A Field Guide to Coastal Wetland Plants of the Northeastern United States by Ralph W. Tiner, Jr., drawings by Abigail Rorer (Amherst: University of Massachusetts Press, 1987), copyright © 1987 by Ralph W. Tiner, Jr. Figures 10 and 17.

Hydric Soils of New England by Ralph W. Tiner, Jr. and Peter L.M. Veneman, drawings by Elizabeth Scott (Amherst: University of Massachusetts Cooperative Extension, 1989). Figure 14.

Acknowledgements

Many individuals have contributed to the completion of the wetlands inventory in Rhode Island and to the preparation of this report. The U.S. Environmental Protection Agency, Region I, Boston contributed funds for publishing this report. Matt Schweisberg served as project officer for this work and his patience is appreciated.

In preparing the National Wetlands Inventory maps, wetland photo interpretation was done by John Organ, Frank Shumway, Judy Harding, and Janice Stone. Their work serves as the foundation for this report. John Organ also provided assistance in quality control of the interpreted photographs and in draft map review. The Service's National Wetlands Inventory Group in St. Petersburg, Florida provided technical support for producing the wetland maps. Nino Alessandroni and Rudolf Nyc were responsible for compiling wetland acreage summaries, while Joanne Gookin also assisted.

The following persons reviewed the draft manuscript and provided comments: Dr. Frank Golet (University of Rhode Island), Bob Scheirer (FWS), Bill Zinni (FWS), Glenn Smith (FWS), Everett Stuart (SCS), Matt Schweisberg (EPA), Rick Enser (RIDEM), Irene Kenenski (RICRMC), and Dean Albro (RIDEM). I am particularly indebted to Dr. Golet who spent considerable time reviewing the manuscript, and preparing comments based on his extensive knowledge of the state's wetlands; he also provided slides for use in this report.

Others providing information included: Porter Reed, Ralph Abele, Charles Allin, Lori Suprock, Michael Lapisky, Tim Lynch, Patricia Toomer, and Alan Steiner.

The manuscript was typed by Joanne Gookin and Joan Gilbert. Some of the figures were prepared by Mary O'Connor. Rick Newton also provided technical support in preparing photographs for publication.

Photo Credits

Photographs used in this report were gratiously provided by several people who are listed below in alphabetical order along with the plate and figure numbers of their photographs: M. Anderson (Figure 22b), Dr. Frank Golet (Plates 6, 9, 11, 12, 13, 14, 15, and 16; Figures 8, 15a, 15b, 18, 19, 20, 21, 24, and 29), Fred Knapp (Figure 22d), Ralph Tiner (Plates 1, 2, 3, 5, 7, 8, and 10; Figures 16 and 30), Dr. Peter Veneman (Plate 4), and Bill Zinni (Figure 22c).

Cover photo credits: Golet (salt marsh, great egret, least bittern, and pickerelweed marsh) and Tiner (inland marsh and turk's-cap lily).

Table of Contents

	Page
Acknowledgements; Photo Credits	iii
Table of Contents	iv
List of Figures	vi
List of Tables	vii
List of Plates	vii

Chapter 1. Introduction 1
   Rhode Island Wetlands Inventory 2
   Description of Study Area 2
   Purpose and Organization of this Report 2
   References 2

Chapter 2. U.S. Fish and Wildlife Service's Wetland Definition and Classification System 4
   Introduction 4
   Wetland Definition 4
   Wetland Classification 7
   References 12

Chapter 3. National Wetlands Inventory Techniques and Results 13
   Introduction 13
   Wetlands Inventory Techniques 13
      Mapping Photography 13
      Photo Interpretation and Collateral Data 13
      Field Investigations 13
      Draft Map Production 15
      Draft Map Review 15
      Final Map Production 15
      Wetland Acreage Compilation 15
   Wetlands Inventory Results 15
      National Wetlands Inventory Maps 15
      Wetland and Deepwater Habitat Acreage Summaries 15
         State Totals 15
         County Totals 17
   Summary 18
   References 18

Chapter 4. Wetland Formation and Hydrology 19
   Introduction 19
   Wetland Formation 19
      Inland Wetland Formation 19
      Coastal Wetland Formation 21
   Wetland Hydrology 23
      Tidal Wetland Hydrology 23
      Nontidal Wetland Hydrology 24
   References 27

Chapter 5. Hydric Soils of Rhode Island 29
   Introduction 29
   Definition of Hydric Soil 29
   Major Categories of Hydric Soils 29
   National List of Hydric Soils 32
   Rhode Island's Hydric Soils 32

County Acreage of Hydric Soils	32
Hydric Soils Descriptions	32
References	35

Chapter 6. Vegetation and Plant Communities of Rhode Island's Wetlands … 36
Introduction … 36
Hydrophyte Definition and Concept … 36
Wetland Plant Communities … 37
    Marine Wetlands … 38
    Estuarine Wetlands … 38
    Riverine Wetlands … 41
    Palustrine Wetlands … 41
    Lacustrine Wetlands … 49
References … 50

Chapter 7. Wetland Values … 52
Introduction … 52
Fish and Wildlife Values … 52
    Fish and Shellfish Habitat … 52
    Waterfowl and Other Bird Habitat … 53
    Mammal and Other Wildlife Habitat … 54
    Rare, Threatened, or Endangered Plants … 55
Environmental Quality Values … 55
    Water Quality Improvement … 55
    Aquatic Productivity … 59
Socio-economic Values … 60
    Flood and Storm Damage Protection … 60
    Shoreline Erosion Control … 62
    Water Supply … 62
    Ground-water Recharge … 63
    Harvest of Natural Products … 63
    Recreation and Aesthetics … 64
Summary … 64
References … 64

Chapter 8. Wetland Protection … 67
Introduction … 67
Wetland Regulation … 67
Wetland Acquisition … 67
Future Actions … 69
References … 70

Appendix: List of Plant Species that Occur in Rhode Island's Wetlands … 71

Enclosure: General Distribution of Rhode Island's Wetlands

List of Tables

No.		Page
1.	Definition of "wetland" according to selected Federal regulations and state statutes	6
2.	Classes and subclasses of wetlands and deepwater habitats	10
3.	Water regime modifiers, both tidal and nontidal groups	11
4.	Salinity modifiers for coastal and inland areas	11
5.	Wetland acreage summaries for Rhode Island	17
6.	Deepwater habitat acreage summaries for Rhode Island	18
7.	Percentage of each county covered by wetland	18
8.	Tidal ranges of mean and spring tides and mean tide level at various locations in Rhode Island	24
9.	Examples of plant indicators of tidal water regimes for Rhode Island's estuarine wetlands	24
10.	Examples of plant indicators of nontidal water regimes for Rhode Island's wetlands	26
11.	Criteria for hydric soils	29
12.	Definitions of the classes of natural soil drainage associated with wetlands	30
13.	Hydric soils of Rhode Island	33
14.	County summaries of hydric soils acreage in Rhode Island	33
15.	Wetland indicator status of various life forms of Rhode Island's wetland plants	36
16.	Examples of four wetland plant types occurring in Rhode Island	37
17.	Examples of palustrine scrub-shrub wetlands in Rhode Island	44
18.	Examples of palustrine forested wetlands in Rhode Island	46
19.	List of major wetland values	52
20.	Plant species of special concern to Rhode Island that occur in wetlands	56
21.	Summary of primary Federal and state laws requiring permits for wetland alteration in Rhode Island	58

List of Plates*

No.
1. Carlisle muck.
2. Scarboro mucky sandy loam.
3. Wareham loamy sand.
4. Ridgebury fine sandy loam (hydric).
5. Ridgebury fine sandy loam (nonhydric).
6. Sudbury sandy loam.
7. Salt marsh in Tiverton.
8. Salt marsh adjacent to Winnapaug Pond, Westerly.
9. Freshwater wetlands in Frying Pan Pond, Richmond.
10. Seasonally flooded marsh along the Chipuxet River, West Kingston.
11. Seasonally flooded marsh on Block Island.
12. Grazed wet meadow in Tiverton.
13. Diamond Bog, Richmond in autumn.
14. Newton Marsh, Westerly.
15. Atlantic white cedar swamp bordering Ell Pond, Rockville.
16. Red maple swamp, Richmond.

---

* Plates lie between pages 38 and 39.

CHAPTER 1.
Introduction

Wetlands are usually periodically flooded lands occurring between uplands and open water bodies such as lakes, rivers, streams, and estuaries. Many wetlands, however, may be isolated from such water bodies. These wetlands are located in areas with seasonally high water tables that are surrounded by upland. Wetlands are commonly referred to by a host of terms based on their location and characteristics, such as salt marsh, tidal marsh, mudflat, wet meadow, cedar swamp, and hardwood swamp. These areas are important natural resources with numerous values, including fish and wildlife habitat, flood protection, erosion control, and water quality maintenance.

The Fish and Wildlife Service (Service) has always recognized the importance of wetlands to waterfowl, other migratory birds and wildlife. The Service's responsibility for protecting these habitats comes largely from international treaties concerning migratory birds and from the Fish and Wildlife Coordination Act. The Service has been active in protecting these resources through various programs. The Service's National Wildlife Refuge System was established to preserve and enhance migratory bird habitat in strategic locations across the country. More than 10 million ducks breed annually in U.S. wetlands and millions more overwinter here. The Service also reviews Federal projects and applications for Federal permits that involve wetland alteration.

Since the 1950's, the Service has been particularly concerned about wetland losses and their impact on fish and wildlife populations. In 1954, the Service conducted its first nationwide wetlands inventory which focused on important waterfowl wetlands. This survey was performed to provide information for considering fish and wildlife impacts in land-use decisions. The results of this inventory were published in a well-known Service report entitled *Wetlands of the United States*, commonly referred to as Circular 39 (Shaw and Fredine 1956).

Since this survey, wetlands have undergone many changes, both natural and human-induced. The conversion of wetlands for agriculture, residential and industrial developments and other uses has continued. During the 1960's, the general public in many states became more aware of wetland values and concerned about wetland losses. They began to realize that wetlands provided significant public benefits besides fish and wildlife habitat, especially flood protection and water quality maintenance. Prior to this time, wetlands were regarded by most people as wastelands, whose best use could only be attained by alteration, e.g., draining for agriculture, dredging and filling for industrial and housing developments and filling with sanitary landfill. Scientific studies demonstrating wetlands values, especially for coastal marshes, were instrumental in increasing public awareness of wetland benefits and stimulating concern for wetland protection. Consequently, several states passed laws to protect coastal wetlands, including Massachusetts (1963), Rhode Island (1965), Connecticut (1969), New Jersey (1970), Maryland (1970), Georgia (1970), New York (1972) and Delaware (1973). Four of these states subsequently adopted inland or nontidal wetland protection legislation: Massachusetts, Rhode Island, Connecticut and New York. Most of the other states in the Nation with coastal wetlands followed the lead of these northeastern states and enacted laws to protect or regulate uses of coastal wetlands. During the early 1970's, the Federal government also assumed greater responsibility for wetlands through Section 404 of the Federal Water Pollution Control Act of 1972 (later amended as the Clean Water Act of 1977) and by strengthening wetland protection under Section 10 of the Rivers and Harbors Act of 1899. Federal permits are now required for many types of construction in many wetlands, although normal agricultural and forestry activities are exempt.

With increased public interest in wetlands and strengthened government regulation, the Service considered how it could contribute to this resource management effort, since it has prime responsibility for protection and management of the Nation's fish and wildlife and their habitats. The Service recognized the need for sound ecological information to make decisions regarding policy, planning, and management of the country's wetland resources, and established the National Wetlands Inventory Project (NWI) in 1974 to fulfill this need. The NWI aims to generate scientific information on the characteristics and extent of the Nation's wetlands. The purpose of this information is to foster wise use of U.S. wetlands and to provide data for making quick and accurate resource decisions.

Two very different kinds of information are needed: (1) detailed maps and (2) status and trends reports. First, detailed wetland maps are needed for impact assessment of site-specific projects. These maps serve a purpose similar to the U.S.D.A. Soil Conservation Service's soil survey maps, the National Oceanic and Atmospheric Administration's coastal and geodetic survey maps, and the U.S. Geological Survey's topographic maps. Detailed wetland maps are used by local, state and Federal agencies as well as by private industry and organizations for many purposes, including watershed management plans, environmental impact assessments, permit reviews, facility and corridor sitings, oil spill contingency plans, natural resource inventories, wildlife surveys and other uses. To date, wetland maps have been prepared for 61% of the lower 48 states, 18% of Alaska, and all of Hawaii. Secondly, national estimates of the current status and recent losses and gains of wetlands are needed in order to provide improved information for reviewing the effectiveness of existing Federal programs and policies, for identifying national or regional problems and for general public awareness. Technical and popular reports about these trends have been recently published (Frayer, et al. 1983; Tiner 1984).

Rhode Island Wetlands Inventory

Rhode Island's wetlands were mapped as part of a larger inventory including Massachusetts, coastal Maine and southern New Hampshire. Although each of these states had mapped wetlands to some extent, there was no consistency from state to state, due to differences in wetland definitions and inventory procedures. The Service's National Wetlands Inventory Project (NWI) has produced a consistent and more up-to-date set of maps and other data for New England wetlands. The Rhode Island wetlands inventory provides government administrators, private industry, and others with improved information for project planning and environmental impact evaluation and for making land-use decisions. This inventory identifies the current status of Rhode Island's wetlands and serves as the base from which future changes can be determined.

Description of the Study Area

Rhode Island is the smallest state in the Nation, occupying 1,058 square miles or 677,120 acres (Rector 1981). The state is divided into five counties: Bristol, Kent, Newport, Providence, and Washington (Figure 1). Narragansett Bay is a dominant feature in the state, as it essentially separates Newport and Bristol Counties from the rest of the state. The landscape is part of the Eastern Deciduous Forest Province, Appalachian Oak Forest Section as defined by Bailey (1978). The northern part of the state falls within the White Pine Region of southern New England as characterized by Bromley (1935).

The climate of Rhode Island has been described by Rector (1981) and elsewhere. In general, the climate is characterized by cold winters and warm summers, with a moderating ocean influence. Average winter temperature is 30°F with lowest temperatures ranging between -10°F and -20°F. Summer temperatures average 70°F and peak in the 90's. The growing season ranges from late March and early April to November. Annual precipitation averages from 44 to 48 inches, with precipitation relatively evenly distributed throughout the year. Thunderstorms occur mostly in the summer. Average snowfall is 36 inches and maximum snowfall usually occurs in February.

Purpose and Organization of this Report

The purpose of this publication is to report the findings of the Service's wetlands inventory of Rhode Island. The discussion will focus on wetlands with a few references to deepwater habitats which were also inventoried. The following chapters will include discussions of wetland concept and classification (Chapter 2), inventory techniques and results (Chapter 3), wetland formation and hydrology (Chapter 4), hydric soils (Chapter 5), wetland vegetation and plant communities (Chapter 6), wetland values (Chapter 7), and wetland protection (Chapter 8). The appendix contains a list of vascular plants associated with Rhode Island's wetlands. Scientific names of plants follow the *National List of Scientific Plant Names* (U.S.D.A. Soil Conservation Service 1982). A figure showing the general distribution of Rhode Island's wetlands and deepwater habitats is provided as an enclosure at the back of this report. (*Note:* This figure shows many forested wetlands of various sizes and only the large emergent and scrub-shrub wetlands, thus many smaller wetlands of these latter types are not depicted; this is perhaps most evident by the lack of wetlands shown for Block Island.)

References

Bailey, R.G. 1978. Description of the Ecoregions of the United States. U.S. Department of Agriculture, Forest Service, Ogden, Utah. 77 pp.

Bromley, S.W. 1935. The original forest types of southern New England. Ecol. Monog. 5(1): 61–89.

Frayer, W.E., T.J. Monahan, D.C. Bowden, and F.A. Graybill. 1983. Status and Trends of Wetlands and Deepwater Habitats in the Conterminous United States, 1950's to 1970's. Dept. of Forest and Wood Sciences, Colorado State University, Ft. Collins. 32 pp.

Rector, D.D. 1981. Soil Survey of Rhode Island. U.S.D.A. Soil Conservation Service, 200 pp. + maps.

Shaw, S.P. and C.G. Fredine. 1956. Wetlands of the United States. Their Extent and Their Value to Waterfowl and other Wildlife, U.S. Fish and Wildlife Service. Circular 39. 67 pp.

Tiner, R.W., Jr. 1984. Wetlands of the United States: Current Status and Recent Trends. U.S. Fish and Wildlife Service, National Wetlands Inventory, Washington, DC. 59 pp.

U.S.D.A. Soil Conservation Service. 1982. National List of Scientific Plant Names. Vol. 1. List of Plant Names. SCS-TP-159. 416 pp.

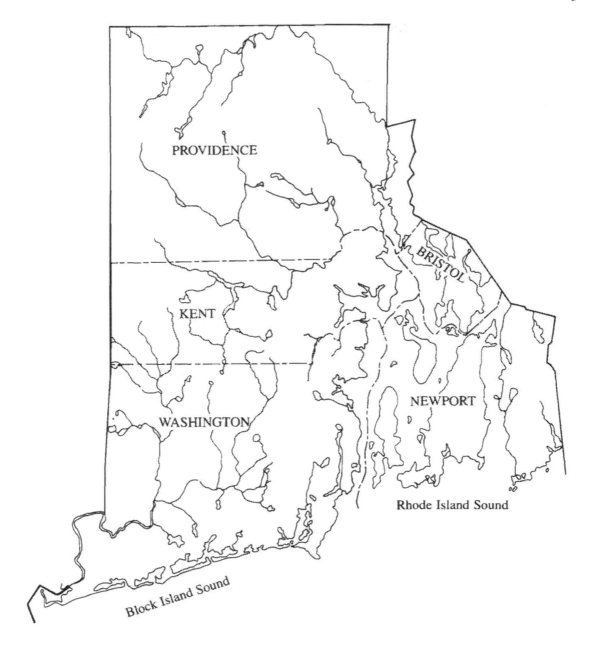

Figure 1. Rhode Island and its counties.

CHAPTER 2.

U.S. Fish and Wildlife Service's Wetland Definition and Classification System

Introduction

To begin inventorying the Nation's wetlands, the Service needed a definition of wetland and a classification system to identify various wetlands types. The Service, therefore, examined recent wetland inventories throughout the country to learn how others defined and classified wetlands. The results of this examination were published as *Existing State and Local Wetlands Surveys (1965-1975)* (U.S. Fish and Wildlife Service 1976). More than 50 wetland classification schemes were identified. Of those, only one classification—the Martin, *et al.* system (1953)—was nationally based, while all others were regionally focused. In January 1975, the Service brought together 14 authors of regional wetland classifications and other prominent wetland scientists to help decide if any existing classification could be used or modified for the national inventory or if a new system was needed. They recommended that the Service attempt to develop a new national wetland classification. In July 1975, the Service sponsored the National Wetland Classification and Inventory Workshop, where more than 150 wetland scientists and mapping experts met to review a preliminary draft of the new wetland classification system. The consensus was that the system should be hierarchical in nature and built around the concept of ecosystems (Sather 1976).

Four key objectives for the new system were established: (1) to develop ecologically similar habitat units, (2) to arrange these units in a system that would facilitate resource management decisions, (3) to furnish units for inventory and mapping, and (4) to provide uniformity in concept and terminology throughout the country (Cowardin, *et al.* 1979).

The Service's wetland classification system was developed by a four-member team, i.e., Dr. Lewis M. Cowardin (U.S. Fish and Wildlife Service), Virginia Carter (U.S. Geological Survey), Dr. Francis C. Golet (University of Rhode Island) and Dr. Edward T. LaRoe (National Oceanic and Atmospheric Administration), with assistance from numerous Federal and state agencies, university scientists, and other interested individuals. The classification system went through three major drafts and extensive field testing prior to its publication as *Classification of Wetlands and Deepwater Habitats of the United States* (Cowardin, *et al.* 1979). Since its publication, the Service's classification system has been widely used by Federal, state, and local agencies, university scientists, and private industry and non-profit organizations for identifying and classifying wetlands. At the First International Wetlands Conference in New Delhi, India, scientists from around the world adopted the Service's wetland definition as an international standard and recommended testing the applicability of the classification system in other areas, especially in the tropics and subtropics (Gopal, *et al.* 1982). Thus, the system appears to be moving quickly towards its goal of providing uniformity in wetland concept and terminology.

Wetland Definition

Conceptually, wetlands usually lie between the better drained, rarely flooded uplands and the permanently flooded deep waters of lakes, rivers and coastal embayments (Figure 2). Wetlands generally include the variety of marshes, bogs, swamps, shallow ponds, and bottomland forests that occur throughout the country. They usually lie in depressions surrounded by upland or along rivers, lakes and coastal waters where they are subject to periodic flooding. Some wetlands, however, occur on slopes where they are associated with ground-water seepage areas. To accurately inventory this resource, the Service had to determine where along this natural wetness continuum wetland ends and upland begins. While many wetlands lie in distinct depressions or basins that are readily observable, the wetland-upland boundary is not always easy to identify. This is especially true along many floodplains, on glacial till deposits, in gently sloping terrain, and in areas of major hydrologic modification. In these areas, only a skilled wetland ecologist or other specialist can accurately identify the wetland boundary. To help ensure accurate and consistent wetland determination, an ecologically based definition was constructed by the Service.

Historically, wetlands were defined by scientists working in specialized fields, such as botany or hydrology. A botanical definition would focus on the plants adapted to flooding or saturated soil conditions, while a hydrologist's defintion would emphasize fluctuations in the position of the water table relative to the ground surface over time. Lefor and Kennard (1977) reviewed numerous definitions for inland wetlands used in the Northeast. Single

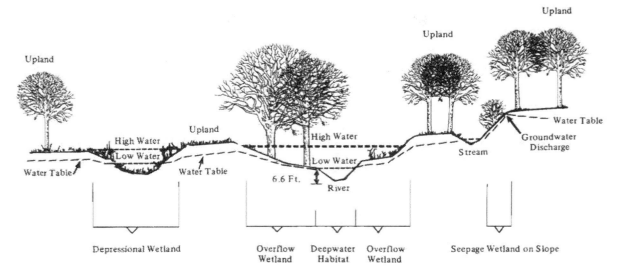

Figure 2. Schematic diagram showing wetlands, deepwater habitats, and uplands on the landscape. Note differences in wetlands due to hydrology and topographic position.

parameter definitions in general are not very useful for identifying wetlands. A more complete definition of wetland involves a multi-disciplinary approach. The Service has taken this approach in developing its wetland definition and classification system.

The Service has not attempted to legally define wetland, since each state or Federal regulatory agency has defined wetland somewhat differently to suit its administrative purposes (Table 1). Therefore, according to existing wetland laws, a wetland is whatever the law says it is. The Service needed a definition that would allow accurate identification and delineation of the Nation's wetlands for resource management purposes.

The Service defines wetlands as follows:

"Wetlands are lands transitional between terrestrial and aquatic systems where the water table is usually at or near the surface or the land is covered by shallow water. For purposes of this classification wetlands must have one or more of the following three attributes: (1) at least periodically, the land supports predominantly hydrophytes; (2) the substrate is predominantly undrained hydric soil; and (3) the substrate is nonsoil and is saturated with water or covered by shallow water at some time during the growing season of each year." (Cowardin, et al. 1979)

In defining wetlands from an ecological standpoint, the Service emphasizes three key attributes of wetlands: (1) hydrology—the degree of flooding or soil saturation, (2) wetland vegetation (hydrophytes), and (3) hydric soils. All areas considered wetland must have enough water at some time during the growing season to stress plants and animals not adapted for life in water or saturated soils. Most wetlands have hydrophytes and hydric soils present, yet many are nonvegetated (e.g., tidal mud flats). The Service has prepared a list of plants occurring in the Nation's wetlands (Reed 1988) and the Soil Conservation Service has developed a national list of hydric soils (U.S.D.A. Soil Conservation Service 1987) to help identify wetlands.

Particular attention should be paid to the reference to flooding or soil saturation during the growing season in the Service's wetland definition. When soils are covered by water or saturated to the surface, free oxygen is generally not available to plant roots. During the growing season, most plant roots must have access to free oxygen for respiration and growth; flooding at this time would have serious implications for the growth and survival of most plants. In a wetland situation, plants must be adapted to cope with these stressful conditions. If, however, flooding only occurs in winter when the plants are dormant, there is little or no effect on them.

Wetlands typically fall within one of the following four categories: (1) areas with both hydrophytes and hydric soils (e.g., marshes, swamps and bogs), (2) areas without hydrophytes, but with hydric soils (e.g., farmed wetlands), (3) areas without soils but with hydrophytes (e.g., seaweed-covered rocky shores), and (4) periodically flooded areas without soil and without hydrophytes (e.g., gravel beaches). All wetlands must be periodcially saturated or covered by shallow water during the growing season, whether or not hydrophytes or hydric soils are present. Completely drained hydric soils that are no

Table 1. Definitions of "wetland" according to selected Federal agencies and state statutes.

Organization (Reference)	Wetland Definition	Comments
U.S. Fish and Wildlife Service (Cowardin, et al. 1979)	"Wetlands are lands transitional between terrestrial and aquatic systems where the water table is usually at or near the surface or the land is covered by shallow water. For purposes of this classification wetlands must have one or more of the following three attributes: (1) at least periodically, the land supports predominantly hydrophytes; (2) the substrate is predominantly undrained hydric soil; and (3) the substrate is nonsoil and is saturated with water or covered by shallow water at some time during the growing season of each year."	This is the official Fish and Wildlife Service definition and is being used for conducting an inventory of the Nation's wetlands. It emphasizes flooding and/or soil saturation, hydric soils and vegetation. Shallow lakes and ponds are included as wetland. Comprehensive lists of wetland plants and soils are available to further clarify this definition.
U.S. Army Corps of Engineers (Federal Register, July 19, 1977) and U.S. Environmental Protection Agency (Federal Register, December 24, 1980)	Wetlands are "those areas that are inundated or saturated by surface or ground water at a frequency and duration sufficient to support, and that under normal circumstances do support, a prevalence of vegetation typically adapted for life in saturated soil conditions. Wetlands generally include swamps, marshes, bogs and similar areas."	Regulatory definition in response to Section 404 of the Clean Water Act of 1977. Excludes similar areas lacking vegetation, such as tidal flats, and does not define lakes, ponds and rivers as wetlands. Aquatic beds are considered "vegetated shallows" and included as other "waters of the United States" for regulatory purposes.
U.S.D.A. Soil Conservation Service (National Food Security Act Manual, 1988)	"Wetlands are defined as areas that have a predominance of hydric soils and that are inundated or saturated by surface or ground water at a frequency and duration sufficient to support, and under normal circumstances do support, a prevalence of hydrophytic vegetation typically adapted for life in saturated soil conditions, except lands in Alaska identified as having a high potential for agricultural development and a predominance of permafrost soils."	This is the Soil Conservation Service's definition for implementing the "Swampbuster" provision of the Food Security Act of 1985. Any area that meets hydric soil criteria is considered to have a predominance of hydric soils. Note the geographical exclusion for certain lands in Alaska.
State of Rhode Island Coastal Resources Mgmt. Council (RI Coastal Resources Mgmt. Program as amended June 28, 1983)	"Coastal wetlands include salt marshes and freshwater or brackish wetlands contiguous to salt marshes. Areas of open water within coastal wetlands are considered a part of the wetland. Salt marshes are areas regularly inundated by salt water through either natural or artificial water courses and where one or more of the following species predominate: [8 indicator plants listed]. Contiguous and associated freshwater or brackish marshes are those where one or more of the following species predominate: [9 indicator plants listed]."	State's public policy on coastal wetlands. Definition based on hydrologic connection to tidal waters and presence of indicator plants. *Note:* Original definition made reference to the occurrence and extent of salt marsh peat; it was probably deleted since many salt marsh soils are not peats, but sands.
State of Rhode Island Dept. of Environmental Mgmt. (RI General Law, Sections 2-1-18 et seq.)	Fresh water wetlands are defined to include, "but not be limited to marshes; swamps; bogs; ponds; river and stream flood plains and banks; areas subject to flooding or storm flowage; emergent and submergent plant communities in any body of fresh water including rivers and streams and that area of land within fifty feet (50') of the edge of any bog, marsh, swamp, or pond." Various wetland types are further defined on the basis of hydrology and indicator plants, including bog (15 types of indicator plants), marsh (21 types of plants), and swamp (24 types of indicator plants plus marsh plants).	Fresh Water Wetlands Act definition. Several wetland types are further defined. The definition includes deepwater areas and the 100-year flood plain as wetland. Minimum size limits are placed on ponds (one quarter acre), marsh (one acre), and swamp (three acres). Under the definition of "river bank," all land within 100 feet of any flowing body of water less than 10 feet wide during normal flow and within 200 feet of any flowing body of water 10 feet or wider is protected as wetland.

longer capable of supporting hydrophytes due to a change in water regime are not considered wetland. Areas with completely drained hydric soils are, however, good indicators of historic wetlands, which may be suitable for restoration through mitigation projects.

It is important to mention that the Service does not generally include permanently flooded deep water areas as wetland, although shallow waters are classified as wetland. Instead, these deeper water bodies are defined as deepwater habitats, since water and not air is the principal medium in which dominant organisms live. Along the coast in tidal areas, the deepwater habitat begins at the extreme spring low tide level. In nontidal freshwater areas, this habitat starts at a depth of 6.6 feet (2 m) because the shallow water areas are often vegetated with emergent wetland plants.

Wetland Classification

The following section represents a simplified overview of the Service's wetland classification system. Consequently, some of the more technical points have been omitted from this discussion. When actually classifying a wetland, the reader is advised to refer to the official classification document (Cowardin, et al. 1979) and should not rely solely on this overview.

The Service's wetland classification system is hierarchial or vertical in nature proceeding from general to specific, as noted in Figure 3. In this approach, wetlands are first defined at a rather broad level—the *SYSTEM*. The term *SYSTEM* represents "a complex of wetlands and deepwater habitats that share the influence of similar hydrologic, geomorphologic, chemical, or biological factors." Five systems are defined: Marine, Estuarine, Riverine, Lacustrine and Palustrine. The Marine System generally consists of the open ocean and its associated high-energy coastline, while the Estuarine System encompasses salt and brackish marshes, nonvegetated tidal shores, and brackish waters of coastal rivers and embayments. Freshwater wetlands and deepwater habitats fall into one of the other three systems: Riverine (rivers and streams), Lacustrine (lakes, reservoirs and large ponds), or Palustrine (e.g., marshes, bogs, swamps and small shallow ponds). Thus, at the most general level, wetlands can be defined as either Marine, Estuarine, Riverine, Lacustrine or Palustrine (Figure 4).

Each system, with the exception of the Palustrine, is further subdivided into *SUBSYSTEMS*. The Marine and Estuarine Systems both have the same two subsystems, which are defined by tidal water levels: (1) Subtidal—continuously submerged areas and (2) Intertidal—areas alternately flooded by tides and exposed to air. Similarly, the Lacustrine System is separated into two systems based on water depth: (1) Littoral—wetlands extending from the lake shore to a depth of 6.6 feet (2 m) below low water or to the extent of nonpersistent emergents (e.g., arrowheads, pickerelweed or spatterdock) if they grow beyond that depth, and (2) Limnetic—deepwater habitats lying beyond the 6.6 feet (2 m) at low water. By contrast, the Riverine System is further defined by four subsystems that represent different reaches of a flowing freshwater or lotic system: (1) Tidal—water levels subject to tidal fluctuations, (2) Lower Perennial—permanent, flowing waters with a well-developed floodplain, (3) Upper Perennial—permanent, flowing water with very little or no floodplain development, and (4) Intermittent—channel containing nontidal flowing water for only part of the year.

The next level—*CLASS*—describes the general appearance of the wetland or deepwater habitat in terms of the dominant vegetative life form or the nature and composition of the substrate, where vegetative cover is less than 30% (Table 2). Of the 11 classes, five refer to areas where vegetation covers 30% or more of the surface: Aquatic Bed, Moss-Lichen Wetland, Emergent Wetland, Scrub-Shrub Wetland and Forested Wetland. The remaining six classes represent areas generally lacking vegetation, where the composition of the substrate and degree of flooding distinguish classes: Rock Bottom, Unconsolidated Bottom, Reef (sedentary invertebrate colony), Streambed, Rocky Shore, and Unconsolidated Shore. Permanently flooded nonvegetated areas are classified as either Rock Bottom or Unconsolidated Bottom, while exposed areas are typed as Streambed, Rocky Shore or Unconsolidated Shore. Invertebrate reefs are found in both permanently flooded and exposed areas.

Each class is further divided into *SUBCLASSES* to better define the type of substrate in nonvegetated areas (e.g., bedrock, rubble, cobble-gravel, mud, sand, and organic) or the type of dominant vegetation (e.g., persistent or nonpersistent emergents, moss, lichen, or broad-leaved deciduous, needle-leaved deciduous, broad-leaved evergreen, needle-leaved evergreen and dead woody plants). Below the subclass level, *DOMINANCE TYPE* can be applied to specify the predominant plant or animal in the wetland community.

To allow better description of a given wetland or deepwater habitat in regard to hydrologic, chemical and soil characteristics and to human impacts, the classification system contains four types of specific modifiers: (1) Water Regime, (2) Water Chemistry, (3) Soil, and (4) Special. These modifiers may be applied to class and lower levels of the classification hierarchy.

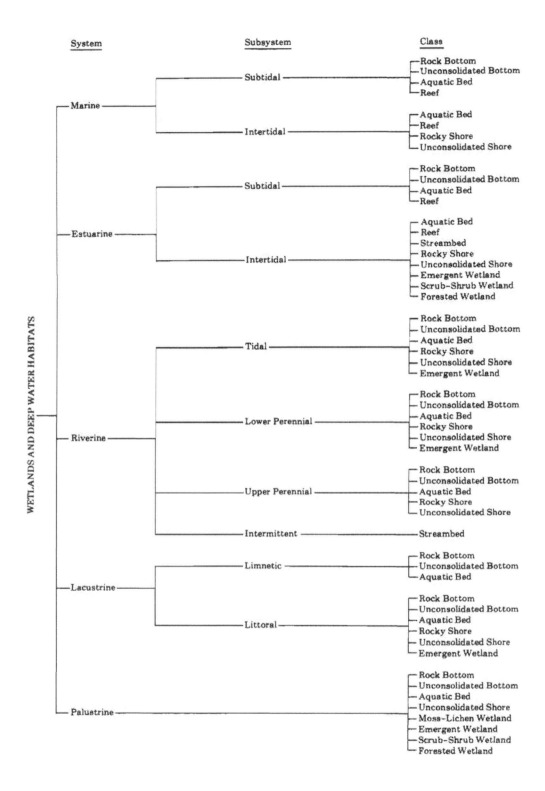

Figure 3. Classification hierarchy of wetlands and deepwater habitats showing systems, subsystems, and classes. The Palustrine System does not include deepwater habitats (Cowardin, *et al.* 1979).

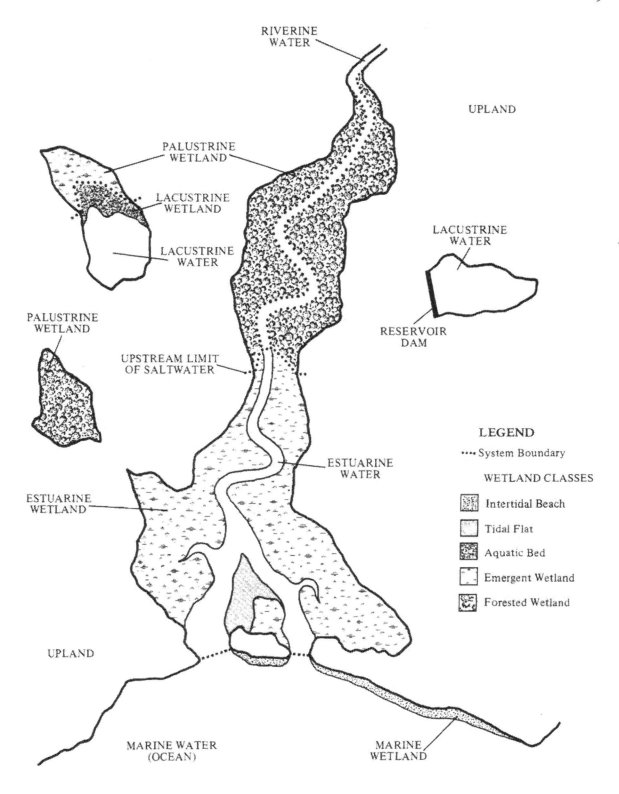

Figure 4. Diagram showing major wetland and deepwater habitat systems. Predominant wetland classes for each system are also designated. *(Note:* Tidal flat and beach classes are now considered unconsolidated shore.)

Table 2. Classes and subclasses of wetlands and deepwater habitats (Cowardin, et al. 1979).

Class	Brief Description	Subclasses
Rock Bottom	Generally permanently flooded areas with bottom substrates consisting of at least 75% stones and boulders and less than 30% vegetative cover.	Bedrock; Rubble.
Unconsolidated Bottom	Generally permanently flooded areas with bottom substrates consisting of at least 25% particles smaller than stones and less than 30% vegetative cover.	Cobble-gravel; Sand; Mud; Organic
Aquatic Bed	Generally permanently flooded areas vegetated by plants growing principally on or below the water surface line.	Algal; Aquatic Moss; Rooted Vascular; Floating Vascular
Reef	Ridge-like or mound-like structures formed by the colonization and growth of sedentary invertebrates.	Coral; Mollusk; Worm
Streambed	Channel whose bottom is completed dewatered at low water periods.	Bedrock; Rubble; Cobble-gravel; Sand; Mud; Organic; Vegetated
Rocky Shore	Wetlands characterized by bedrock, stones or boulders with areal coverage of 75% or more and with less than 30% coverage by vegetation.	Bedrock; Rubble
Unconsolidated Shore*	Wetlands having unconsolidated substrates with less than 75% coverage by stone, boulders and bedrock and less than 30% vegetative cover, except by pioneer plants.	Cobble-gravel; Sand; Mud; Organic; Vegetated
	(*NOTE: This class combines two classes of the 1977 operational draft system—Beach/Bar and Flat)	
Moss-Lichen Wetland	Wetlands dominated by mosses or lichens where other plants have less than 30% coverage.	Moss; Lichen
Emergent Wetland	Wetlands dominated by erect, rooted, herbaceous hydrophytes.	Persistent; Nonpersistent
Scrub-Shrub Wetland	Wetlands dominated by woody vegetation less than 20 feet (6 m) tall.	Broad-leaved Deciduous; Needle-leaved Deciduous; Broad-leaved Evergreen; Needle-leaved Evergreen; Dead
Forested Wetland	Wetlands dominated by wood vegetation 20 feet (6 m) or taller.	Broad-leaved Deciduous; Needle-leaved Deciduous; Broad-leaved Evergreen; Needle-leaved Evergreen; Dead

Water regime modifiers describe flooding or soil saturation conditions and are divided into two main groups: (1) tidal and (2) nontidal. Tidal water regimes are used where water level fluctuations are largely driven by oceanic tides. Tidal regimes can be subdivided into two general categories, one for salt and brackish water tidal areas and another for freshwater tidal areas. This distinction is needed because of the special importance of seasonal river overflow and ground-water inflows in freshwater tidal areas. By contrast, nontidal modifiers define conditions where surface water runoff, ground-water discharge, and/or wind effects (i.e., lake seiches) cause water level changes. Both tidal and nontidal water regime modifiers are presented and briefly defined in Table 3.

Water chemistry modifiers are divided into two categories which describe the water's salinity or hydrogen ion concentration (pH): (1) salinity modifiers and (2) pH modifiers. Like water regimes, salinity modifiers have been further subdivided into two groups: halinity modifiers for tidal areas and salinity modifiers for nontidal areas. Estuarine and marine waters are dominated by sodium chloride, which is gradually diluted by fresh water as one moves upstream in coastal rivers. On the other hand, the salinity of inland waters is dominated by four major cations (i.e., calcium, magnesium, sodium and potassium) and three major anions (i.e., carbonate, sulfate, and chloride). Interactions between precipitation, surface runoff, ground-water flow, evaporation, and sometimes plant evapotranspiration form inland salts which are most common in arid and semiarid regions of the country. Table 4 shows ranges of halinity and salinity modifiers which are a modification of the Venice System (Remane and Schlieper 1971). The other set of water chemistry modifiers are pH modifiers for identifying acid (pH<5.5), circumneutral (5.5–7.4) and alkaline (pH>7.4) waters. Some studies have shown a good correlation between plant distribution and pH levels (Sjors 1950; Jeglum 1971). Moreover, pH can be used to distinguish between mineral-rich (e.g., fens) and mineral-poor wetlands (e.g., bogs).

The third group of modifiers—soil modifiers—are presented because the nature of the soil exerts strong influ-

Table 3. Water regime modifiers, both tidal and nontidal groups (Cowardin, et al. 1979).

Group	Type of Water	Water Regime	Definition
Tidal	Saltwater and brackish areas	Subtidal	Permanently flooded tidal waters
		Irregularly exposed	Exposed less often than daily by tides
		Regularly flooded	Daily tidal flooding and exposure to air
		Irregularly flooded	Flooded less often than daily and typically exposed to air
	Freshwater	Permanently flooded-tidal	Permanently flooded by tides and river or exposed irregularly by tides
		Semipermanently flooded-tidal	Flooded for most of the growing season by river overflow but with tidal fluctuation in water levels
		Regularly flooded	Daily tidal flooding and exposure to air
		Seasonally flooded-tidal	Flooded irregularly by tides and seasonally by river overflow
		Temporarily flooded-tidal	Flooded irregularly by tides and for brief periods during growing season by river overflow
Nontidal	Inland freshwater and saline areas	Permanently flooded	Flooded throughout the year in all years
		Intermittently exposed	Flooded year-round except during extreme droughts
		Semipermanently flooded	Flooded throughout the growing season in most years
		Seasonally flooded	Flooded for extended periods in growing season, but surface water is usually absent by end of growing season
		Saturated	Surface water is seldom present, but substrate is saturated to the surface for most of the season
		Temporarily flooded	Flooded for only brief periods during growing season, with water table usually well below the soil surface for most of the season
		Intermittently flooded	Substrate is usually exposed and only flooded for variable periods without detectable seasonal periodicity (Not always wetland: may be upland in some situations)
		Artificially flooded	Duration and amount of flooding is controlled by means of pumps or siphons in combination with dikes or dams

Table 4. Salinity modifiers for coastal and inland areas (Cowardin et al., 1979).

Coastal Modifiers[1]	Inland Modifiers[2]	Salinity (‰)	Approximate Specific Conductance (Mhos at 25° C)
Hyperhaline	Hypersaline	>40	>60,000
Euhaline	Eusaline	30–40	45,000–60,000
Mixohaline (Brackish)	Mixosaline[3]	0.5–30	800–45,000
Polyhaline	Polysaline	18–30	30,000–45,000
Mesohaline	Mesosaline	5–18	8,000–30,000
Oligohaline	Oligosaline	0.5–5	800–8,000
Fresh	Fresh	<0.5	<800

[1]Coastal modifiers are employed in the Marine and Estuarine Systems.
[2]Inland modifiers are employed in the Riverine, Lacustrine and Palustrine Systems.
[3]The term "brackish" should not be used for inland wetlands or deepwater habitats.

ences on plant growth and reproduction as well as on the animals living in it. Two soil modifiers are given: (1) mineral and (2) organic. In general, if a soil has 20 percent or more organic matter by weight in the upper 16 inches, it is considered an organic soil, whereas if it has less than this amount, it is a mineral soil. For specific definitions, please refer to Appendix D of the Service's classification system (Cowardin, et al. 1979) or to *Soil Taxonomy* (Soil Survey Staff 1975).

The final set of modifiers—special modifiers—were established to describe the activities of people or beaver affecting wetlands and deepwater habitats. These modifiers include: excavated, impounded (i.e., to obstruct outflow of water), diked (i.e., to obstruct inflow of water), partly drained, farmed, and artificial (i.e., materials deposited to create or modify a wetland or deepwater habitat).

References

Cowardin, L.M., V. Carter, F.C. Golet and E.T. LaRoe. 1977. Classification of Wetlands and Deep-water Habitats of the United States (An Operational Draft). U.S. Fish and Wildlife Service. October 1977. 100 pp.

Cowardin, L.M., V. Carter, F.C. Golet and E.T. LaRoe. 1979. Classification of Wetlands and Deepwater Habitats of the United States. U.S. Fish and Wildlife Service, Washington, DC. FWS/OBS-79/31. 103 pp.

Gopal, B., R.E. Turner, R.G. Wetzel and D.F. Whigham. 1982. Wetlands Ecology and Management. Proceedings of the First International Wetlands Conference (September 10–17, 1980; New Delhi, India). National Institute of Ecology and International Scientific Publications, Jaipur, India. 514 pp.

Jeglum, J.K. 1971. Plant indicators of pH and water level in peat lands at Candle Lake, Saskatchewan. Can. J. Bot. 49: 1661–1676.

Lefor, M.W. and W.C. Kennard. 1977. Inland Wetland Definitions. University of Connecticut, Institute of Water Resources, Storrs. Report No. 28. 63 pp.

Martin, A.C., N. Hotchkiss, F.M. Uhler and W.S. Bourn. 1953. Classification of Wetlands of the United States. U.S. Fish and Wildlife Service, Washington, DC. Special Scientific Report, Wildlife No. 20. 14 pp.

Reed, P.B., Jr. 1988. National List of Plant Species that Occur in Wetlands: 1988 National Summary. U.S. Fish and Wildlife Service, National Ecology Research Center, Ft. Collins, CO. Biol. Rep. 88(24). 244 pp.

Remane, A. and C. Schlieper. 1971. Biology of Brackish Water. Wiley Interscience Division, John Wiley & Sons, New York. 372 pp.

Sather, J.H. (editor). 1976. Proceedings of the National Wetland Classification and Inventory Workshop, July 20–23, 1975, at the University of Maryland. U.S. Fish and Wildlife Service, Washington, DC. 358 pp.

Shaw, S.P. and C.G. Fredine. 1956. Wetlands of the United States. U.S. Fish and Wildlife Service, Washington, DC. Circular 39. 67 pp.

Sjors, H. 1950. On the relation between vegetation and electrolytes in north Swedish mire waters. Oikos 2: 241–258.

Soil Survey Staff. 1975. Soil Taxonomy. Department of Agriculture, Soil Conservation Service, Washington, DC. Agriculture Handbook No. 436. 754 pp.

U.S. Fish and Wildlife Service. 1976. Existing State and Local Wetlands Surveys (1965–1975). Volume II. Narrative. Office of Biological Services, Washington, DC. 453 pp.

U.S.D.A. Soil Conservation Service. 1987. Hydric Soils of the United States. In cooperation with the National Technical Committee for Hydric Soils. Washington, DC.

CHAPTER 3.
National Wetlands Inventory Mapping Techniques and Results

Introduction

The National Wetlands Inventory Project (NWI) utilizes remote sensing techniques with supplemental field investigations for wetland identification and mapping. High-altitude aerial photography ranging in scale from 1:58,000 to 1:80,000 serves as the primary remote imagery source. Once suitable high-altitude photography is obtained, there are seven major steps in preparing wetland maps: (1) field investigations, (2) photo interpretation, (3) review of existing wetland information, (4) quality assurance, (5) draft map production, (6) interagency review of draft maps, and (7) final map production. Steps 1, 2 and 3 encompass the basic data collection phase of the inventory. After publication of final wetland maps for Rhode Island, the Service began collecting acreage data on the state's wetlands and deepwater habitats. The procedures used to inventory Rhode Island's wetlands and the results of this inventory are discussed in the following sections.

Wetlands Inventory Techniques

Mapping Photography

For mapping Rhode Island's wetlands, the Service used 1:80,000 black and white photography (Figure 5). Most of this imagery was acquired from the spring of 1974 to the spring of 1977. Thus, the effective period of this inventory can be considered the mid-1970's.

Photo Interpretation and Collateral Data

Photo interpretation was performed by the Department of Forestry and Wildlife Management, University of Massachusetts, Amherst. All photo interpretation was done in stereo using mirror stereoscopes. Other collateral data sources used to aid in wetland detection and classification included:

(1) 1:20,000 black and white photography (1975);
(2) U.S. Geological Survey topographic maps;
(3) U.S.D.A. Soil Conservation Service soil surveys;
(4) U.S. Department of Commerce coastal and geodetic survey maps;
(5) Rhode Island Map Down Land Use and Vegetative Cover Maps (MacConnell 1974).

Wetland photo interpretation, although extremely efficient and accurate for inventorying wetlands, does have certain limitations. Consequently, some problems arose during the course of the survey. Additional field work or use of collateral data was necessary to help overcome these constraints. These problems and their resolution are discussed below.

1. Spring flooding of certain wetlands. In some areas, spring flooding produced a dark photo signature, obscuring the wetland vegetation. Field checks and use of collateral photography allowed determination of appropriate vegetation class.

2. Identification of freshwater aquatic beds and nonpersistent emergent wetlands. Due to the primary use of spring photography, these wetland types were not interpretable. They were generally classified as open water, unless vegetation was observed during field investigations.

3. Identification of subclass in forested wetlands. Due to the spring flooding problem, field checks had to be conducted in many areas to distinguish deciduous from evergreen trees.

4. Inclusion of small upland areas within delineated wetlands. Small islands of higher elevation and better drained uplands naturally exist within many wetlands. Due to the minimum size of mapping units, small upland areas may be included within designated wetlands. Field inspections and/or use of larger-scale photography were used to refine wetland boundaries when necessary.

5. Forested wetlands on glacial till. These wetlands are difficult to identify in the field, let alone through air photo interpretation. Consequently, some of these wetlands were not detected and do not appear on the NWI maps.

Field Investigations

Ground truthing surveys were conducted to collect information on plant communities of various wetlands and to gain confidence in detecting and classifying wetlands from aerial photography. Detailed notes were taken at more than 70 sites throughout the state. In addition to

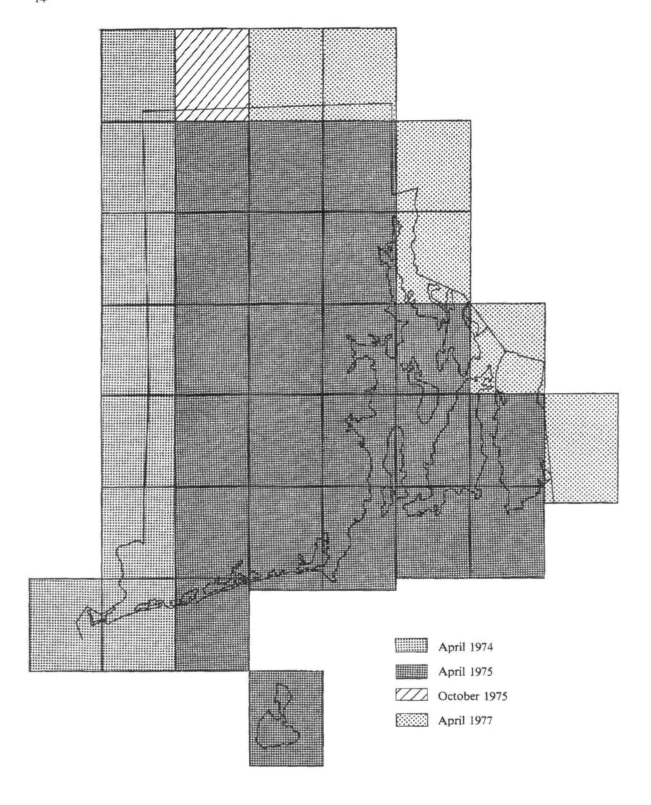

Figure 5. Index of aerial photography used for the National Wetlands Inventory in Rhode Island

these sites, observations were made at countless other wetlands for classification purposes and notations were recorded on appropriate topographic maps. In total, approximately three weeks were spent in the field examining wetlands.

Draft Map Production

Upon completion of photo interpretation, two levels of quality assurance were performed: (1) regional quality control, and (2) national consistency quality assurance. Regional review of each interpreted photo was accomplished by Regional Office's NWI staff to ensure identification of all wetlands and proper classification. By contrast, national quality control by the NWI Group at St. Petersburg, Florida entailed spot checking of photos to ensure that national standards had been successfully followed. Once approved by quality assurance, draft large-scale (1:24,000) wetland maps were produced by the Group's support service contractor using Bausch and Lomb zoom transfer scopes.

Draft Map Review

Draft maps were sent to the following agencies for review and comment:

(1) U.S. Fish and Wildlife Service, New England Field Office;
(2) U.S. Army Corps of Engineers (New England Division);
(3) U.S.D.A. Soil Conservation Service;
(4) U.S. Environmental Protection Agency (Region I);
(5) National Marine Fisheries Service; and
(6) Rhode Island Department of Environmental Management.

In addition, the Regional Office's NWI staff conducted field checks and a thorough examination of draft maps to ensure proper placement of wetland polygons and labels as well as accurate classification.

Final Map Production

All comments received were evaluated and incorporated into the final maps, as appropriate. Final maps were published in 1980.

Wetland Acreage Compilation

In 1984, the Service collected wetland acreage information from NWI maps for Rhode Island. Area measurements were recorded using a video area measurement system (VAMS) rather than conventional planimeters. The VAMS is a photo-optic system that allows the operator to fill in map polygons electronically, while the acreage is automatically determined and then recorded in a computer file. This technique allowed for preparation of acreage summaries on a county basis and for the entire state.

Wetlands Inventory Results

National Wetlands Inventory Maps

A total of thirty-seven (37) 1:24,000-scale wetland maps were produced. These maps identify the size, shape and type of wetlands and deepwater habitats in accordance with NWI specifications. The minimum mapping unit for wetlands ranges between approximately 1–3 acres. A recent evaluation of NWI maps in Massachusetts determined that these maps had accuracies exceeding 95% (Swartwout, *et al.* 1982). This high accuracy is possible because the inventory technique involves a combination of photo interpretation, field studies, use of existing information and interagency review of draft maps. Final maps have been available since 1980. Figure 6 shows an example of the large-scale map. In the near future, a series of small-scale wetland maps (1:100,000) will be produced by the NWI. Copies of NWI maps and a map catalogue can be ordered from the Rhode Island Department of Environmental Management, Freshwater Wetlands Section, 83 Park Street, Providence, RI 02903 (401/277-6820).

Wetland and Deepwater Habitat Acreage Summaries

State Totals

According to this survey, Rhode Island possesses roughly 65,000 acres of wetlands and 106,000 acres of deepwater habitats, excluding marine waters and smaller rivers and streams that either appear as linear features on wetland maps or wetlands that were not identified due to their small size. About 10 percent of the state's land surface is represented by wetlands.

Nearly all (99 percent) of the state's wetlands fall within two systems—palustrine (88 percent) and estuarine (11 percent). The general distribution of Rhode Island's wetlands and deepwater habitats is shown on the enclosed figure at the back of this report.

Palustrine wetlands are slightly more than eight times more abundant than estuarine wetlands, covering 57,106 acres. Eighty-three percent of Rhode Island's freshwater wetlands are forested wetlands, with 77 percent of the total represented by broad-leaved deciduous forested wet-

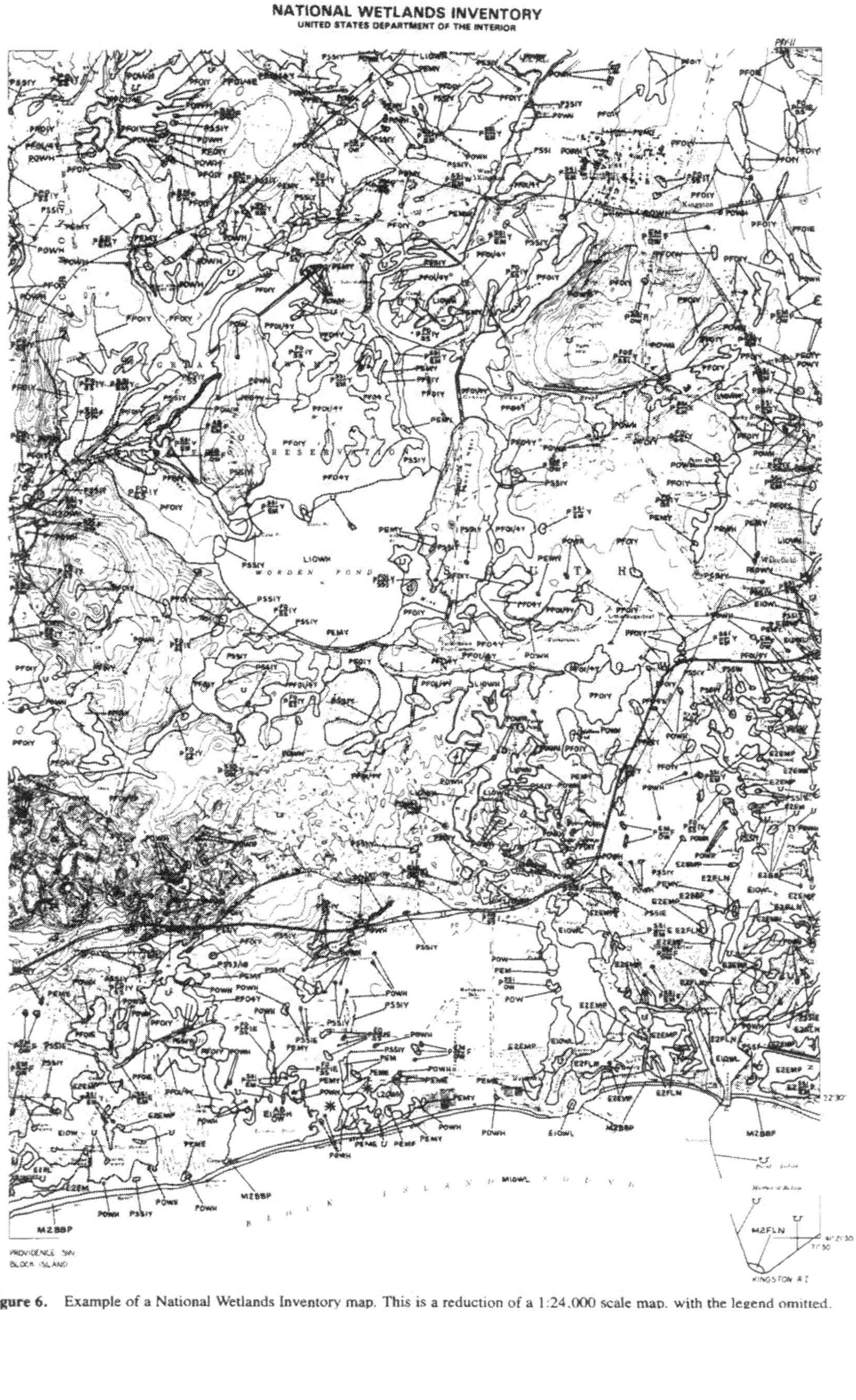

Figure 6. Example of a National Wetlands Inventory map. This is a reduction of a 1:24,000 scale map, with the legend omitted.

Table 5. Wetland acreage summaries for Rhode Island.

System	Class	Bristol County	Kent County	Newport County	Providence County	Washington County	State Total
Marine	Beach/Bar	—	—	147	—	342	489
	Rocky Shore	—	—	217	—	122	339
	Flat	—	—	99	—	4	103
	(Subtotal)	—	—	(463)	—	(468)	(931)
Estuarine	Beach/Bar	125	16	366	6	64	577
	Rocky Shore	—	—	20	—	2	22
	Flat	257	148	401	166	1,927	2,899
	Emergent Wetland	820	164	1,081	72	1,331	3,468
	Scrub-Shrub Wetland	—	—	—	—	52	52
	(Subtotal)	(1,202)	(328)	(1,868)	(244)	(3,376)	(7,018)
Palustrine	Open Water	32	389	302	783	929	2,435
	Aquatic Bed	1	17	1	14	41	74
	Emergent Wetland	33	180	378	691	397	1,679
	Scrub-Shrub Wetland						
	Deciduous	103	914	635	1,205	2,326	5,183
	Evergreen	—	7	—	54	58	119
	Forested Wetland						
	Broad-leaved Deciduous	658	5,880	4,359	12,032	21,219	44,148
	Needle-leaved Evergreen	—	566	3	820	1,850	3,239
	Dead	—	19	—	46	80	145
	Farmed Wetland (Cranberry Bog)	—	74	—	—	10	84
	(Subtotal)	(827)	(8,046)	(5,678)	(15,645)	(26,910)	(57,106)
Lacustrine	Aquatic Bed/Emergent Wetland (nonpersistent)	—	—	—	38	61	99
Total Wetlands		2,029	8,374	8,009	15,927	30,815	65,154

lands, mostly red maple swamps. Evergreen forested wetlands accounted for nearly six percent, whereas scrub-shrub wetlands make up nine percent of the freshwater wetlands. By contrast, emergent wetlands represent only three percent of the total inland wetlands, while small ponds (including aquatic beds) made up the remaining four percent.

Of the 7,018 estuarine wetland acres, 49 percent are emergent wetlands and 41 percent are intertidal flats. Scrub-shrub wetlands account for only 52 acres, whereas estuarine beaches total 577 acres.

Deepwater habitats in Rhode Island, excluding marine waters, total 106,257 acres. As expected, due to Narragansett Bay, estuarine waters predominate and represent 82 percent of the total. Lakes and reservoirs accounted for about 18 percent, while freshwater rivers only made up 0.5 percent of the state's non-marine waters.

County Totals

Acreage of wetlands and deepwater habitats for each county are presented in detail on Tables 5 and 6, respectively. Table 7 shows the percentage of each county covered by wetlands. Washington County possesses 30,815 acres or 47 percent of the state's wetlands, while Providence County has 15,927 acres of wetland or nearly 25 percent of the state total. Newport and Kent Counties have nearly equal amounts of wetland and their combined total represented another 25 percent. Bristol County has 2,029 acres or only three percent of the state's wetlands.

Marine wetlands, largely represented by intertidal beaches and rocky shores, occur only in Washington and Newport Counties, whereas estuarine and palustrine wetlands exist in all five counties. Nearly half (48 percent or 3,376 acres) of the state's estuarine wetlands are found in Washington County, while 27 percent (or 1,868 acres) occur in Newport County. Washington County also possesses 26,910 acres of palustrine wetlands which account for 47 percent of the state's freshwater wetlands. Providence County has 15,838 acres of palustrine wetlands or 28 percent of the state total. Lacustrine wetlands are rather limited, and were mapped only in Washington and Providence Counties.

Newport County has the greatest acreage of non-marine deepwater habitats, with 50,137 acres inventoried. This

Table 6. Deepwater habitat acreage summaries for Rhode Island, excluding marine waters.

System	Bristol County	Kent County	Newport County	Providence County	Washington County	State Total
Estuarine	9,687	6,668	48,254	3,560	18,740	86,909
Riverine	—	71	—	312	143	526
Lacustrine	160	2,216	1,883	10,075	4,488	18,822
Totals	9,847	8,955	50,137	13,947	23,371	106,257

figure accounts for 47 percent of the state's non-marine waters. Washington County possesses 23,371 acres or 22 percent of the state total.

Summary

The NWI Project has completed an inventory of Rhode Island's wetlands using aerial photo interpretation methods. Detailed wetland maps and acreage summaries have been produced for the entire state. Roughly 65,000 acres of wetland and 106,000 acres of deepwater habitat were inventoried in Rhode Island. Thus, about 10 percent of the state is represented by wetland.

Table 7. Percentage of each county covered by wetland. County acreages were obtained from U.S.D.A. soil survey for Rhode Island (1981).

County	County Acreage	Wetland Acreage	% of County Covered by Wetland
Bristol	16,090	1,836	11.4
Kent	111,410	8,374	7.5
Newport	67,740	8,009	11.8
Providence	268,130	16,120	6.0
Washington	213,750	30,815	14.4

References

Cowardin, L.M., V. Carter, F.C. Golet and E.T. LaRoe. 1977. Classification of Wetlands and Deep-water Habitats of the United States (An Operational Draft). U.S. Fish and Wildlife Service. October 1977. 100 pp.

Cowardin, L.M., V. Carter, F.C. Golet and E.T. LaRoe. 1979. Classification of Wetlands and Deepwater Habitats of the United States. U.S. Fish and Wildlife Service. FWS/OBS-79/31. 103 pp.

MacConnell, W.P. 1974. Remote Sensing Land Use and Vegetative Cover in Rhode Island. Cooperative Extension Service, University of Rhode Island, Kingston. 93 pp.

Swartwout, D.J., W.P. MacConnell, and J.T. Finn. 1982. An Evaluation of the National Wetlands Inventory in Massachusetts. Proc. of In-Place Resource Inventories Workshop (University of Maine, Orono, August 9–14, 1981). pp. 658–691.

U.S.D.A. Soil Conservation Service. 1981. Soil Survey of Rhode Island. 200 pp. + maps.

CHAPTER 4.

Wetland Formation and Hydrology

Introduction

Wetlands are usually found in depressions, along the shores of waterbodies such as lakes, rivers, coastal ponds, and estuaries, and at the toes of slopes. Some wetlands occur on the slopes themselves where they are associated with groundwater seepage (springs) or with surface water drainageways. Historical events and present hydrologic conditions have acted in concert to create and maintain a diversity of wetlands in Rhode Island. Human activities have recently become more important to wetland formation and hydrology. The following sections address general differences between Rhode Island's inland and coastal wetlands in terms of their formation and hydrology. The discussion is not intended to be comprehensive, but a generalized overview of wetland formation and hydrology. References have been cited for more detailed descriptions.

Wetland Formation

Many events have led to the creation of wetlands throughout the state, but none are more significant than glaciation, a geological event that took place thousands of years ago. Current events, such as rising sea level and erosion and accretion processes, continue to build, shape, and even destroy wetlands. Construction of ponds, impoundments, and reservoirs also may create wetlands, but often involve wetland destruction as well.

Inland Wetland Formation

Glaciation was the most important factor in the creation of Rhode Island's inland wetlands. From about one billion years ago to about 10–12,000 years ago, the state was buried under several thousand feet of glacial ice (Quinn 1973). During this "ice age," roughly one-third of the world's land area was covered by ice compared to only ten percent of today's surface (Wolfe 1977). During the last glaciation (Wisconsinan), glacial ice advanced as far south as northern New Jersey and northeastern Pennsylvania in the East and to southern Illinois in the Midwest (Shepps 1978).

As the glacier moved south, the landscape was scoured and in most areas the soil was carried away by the moving ice, leaving mostly fresh bedrock behind. In some areas, depressions were scraped out of the bedrock. When the glacier stopped moving south, the leading edge of the ice melted, depositing boulders, sand, and clay (glacial till) and forming an irregular ridge called an "end moraine." The outermost end moraine in Rhode Island is located offshore on Block Island; it extends west to the middle of Long Island and east to Martha's Vineyard and Nantucket (Quinn 1973).

When the climate warmed and melting increased about 18,000 years ago, the glacier retreated northward to a point just north of Rhode Island's South Shore. There it stopped for several thousand years, depositing glacially held materials and forming another end moraine, the Charlestown Moraine. This ridge promoted wetland formation by blocking drainages from the north and causing the newly formed Pawcatuck River to flow west instead of south. When the glacier finally retreated from Rhode Island, drainage patterns throughout the state were modified greatly. Glacial drift was dumped across the landscape—till at higher elevations and stratified drift in lowlands (F. Golet, pers. comm.). Other areas were deeply eroded. When this drift blocked outlets of streams, wetlands and waterbodies were created. As the glacier melted, ice blocks of assorted sizes were left in many places on the outwash plains in front of the melting glacier where they were buried by outwash or surrounded by it. Ice blocks were also deposited in till areas and on the end moraine itself. Glacial ice blocks gradually melted and many formed kettle lakes and ponds (Figure 7). Many of these waterbodies gradually filled with sediment from inflowing streams. Eventually, aquatic vegetation colonized the shallow waters and began to flourish, building peat deposits, and increasing the level of the wetland surface, so that other wetland plants could become established. Filling of glacial lakes and ponds still continues (Figure 8), yet some isolated ones on the Charlestown moraine are still deep—greater than 50 feet (F. Golet, pers. comm.). Many of Rhode Island's wetlands were formed in these types of glacial depressions. Dansereau and Segadas-Vianna (1952) have described the development of eastern North American bogs in these situations (Figure 9).

Wetlands also developed in compact glacial till soils with a confining layer (hardpan or fragipan) close to the surface. In certain soils (e.g., Ridgebury and Stissing series) slowly permeable layers form due to the downward migration (translocation) of silts, clays, and iron oxides and their accumulation in a distinct zone or to

Figure 7. Many inland wetlands have developed in kettles, while coastal wetlands have formed behind barrier beaches, along tidal rivers and in sheltered coves. The region shown is in Washington County (Point Judith - lower right, Wakefield/Pettaquamscutt Cove - upper right, Worden Pond - upper left, Trustom and Card Ponds - lower left). Area A shows kettle ponds in South Kingston; Area B - coastal marshes and tidal flats in Galilee and Jerusalem; Area C - coastal marsh along Pettaquamscutt River.

compaction of till by past glaciers. When close to the surface (e.g., within 20 inches) these layers can create "perched" water tables that promote the establishment of wetland plants and the formation of wetlands. These soils occur in depressions, but also are found on gentle slopes in glacial till areas.

Wetlands have formed on floodplains along certain large streams in the state. In mature floodplains, wetlands are found on the inner floodplain terrace behind the natural levees. The levees are composed of coarser materials and are better drained than the inner floodplain that is composed of silts and clays and has generally poor drainage. Early stages of floodplain development are characterized by extensive marshes bordering streams, while later stages developing as sedimentation increases wetland surface elevations are usually represented by shrub and forested wetlands (Nichols 1915). Some floodplain marshes and meadows may persist due to extended flooding periods that preclude the establishment of tree species or to periodic mowing or grazing.

Figure 8. Aerial view of Hannah Clarkin Pond (a kettle pond) in Charlestown. Note the floating aquatic beds and Atlantic white cedar stands along the water's edge.

Historically, the activities of beaver were instrumental in creating wetlands. Through their construction of dams, beaver blocked drainages, causing water levels to rise and flood adjacent uplands. Flooding of uplands killed the existing vegetation and allowed wetland plants to become established. Although extirpated from the state due to fur trapping in colonial times, beaver populations are now increasing, especially in the Moosup River and Pawcatuck systems in western and southern Rhode Island.

Human activities have become increasingly important in wetland creation. Construction of farm ponds, sedimentation/detention ponds, recreational ponds and lakes, and reservoirs may create vegetated wetlands to some extent, although natural wetlands may be altered or destroyed by these projects. Farm ponds, for example, may become overgrown with wetland vegetation including aquatic plants and emergent, herbaceous plants. Shrub and forested wetlands could eventually become established in manmade basins. Wetland vegetation may also develop along the shorelines of the larger waterbodies. Unfortunately, reservoirs are usually subjected to drastic drawdowns of water levels during the summer leaving exposed, nonvegetated shores which are unsuitable for establishment of a viable wetland plant community. More stable water levels would, however, promote wetland formation along shorelines. Wetlands have been unintentionally created in some areas by highways that directly block former drainageways or that have undersized culverts causing a local rise in water levels. In other cases, wetlands may be intentionally created to mitigate unavoidable losses of natural wetlands by various construction projects.

Coastal Wetland Formation

Coastal wetlands have also been affected by glaciation, but in a much different way than their inland counterparts. During the "ice age," much of the world's ocean waters were stored in the form of glacial ice. At that time, sea level was as much as 425 feet lower than present levels (Wolfe 1977). As the glaciers melted (deglaciation), water was released back into the oceans, thereby raising sea levels. As sea level rose, barrier islands migrated landward and river valleys were submerged. Coastal marshes which were behind these barrier islands were submerged along with other low-lying areas, but other coastal wetlands eventually reformed behind the barrier beaches when they finally stabilized. Evidence of submergence of low-lying areas may be found in salt marshes where buried Atlantic white cedar stumps may be present, e.g., Hundred Acre Cove in Barrington (F. Golet, pers. comm.). Rising sea level has transformed former nontidal freshwater wetlands and uplands into coastal marshes. Bloom and Ellis (1965) reported on the occurrence of salt marshes over former

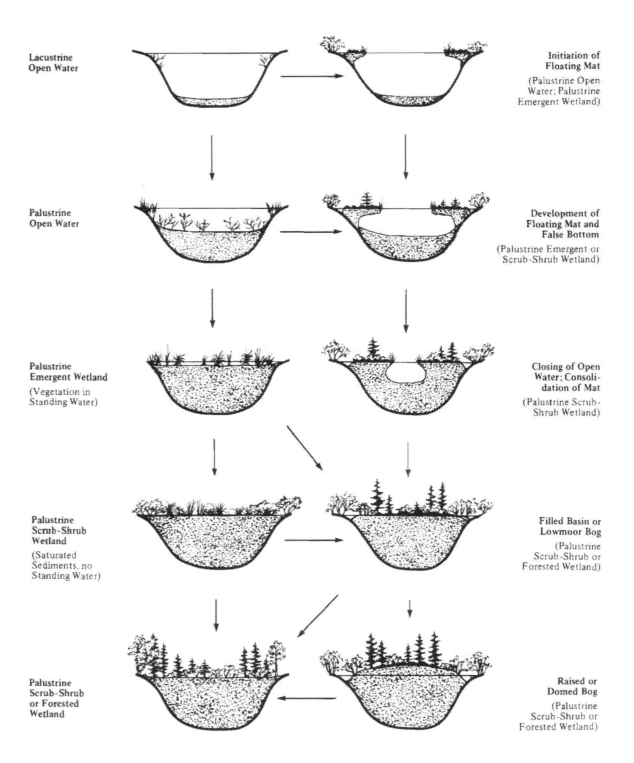

Figure 9. Marsh and bog successional patterns in kettle depressions (adapted from Dansereau and Segadas-Vianna 1952)

uplands in Connecticut. Narragansett Bay is an excellent example of a "drowned" river valley. Coastal wetlands have formed in various coves or protected embayments along the Bay.

Today, sea level continues to rise along the U.S. coastline at average rates between four and ten inches per century, with local variations (Hicks, et al. 1983). The "greenhouse effect" and projected global warming could lead to further melting of polar ice in Greenland and the Antarctic and of mountain glaciers; this coupled with coastal subsidence could raise sea levels 3.0 to 5.7 feet (3.7 feet most likely) by the year 2100 (Titus and Seidel 1986). Such an increase would have profound effects on Rhode Island's coastal wetlands as well as other low-lying areas in the coastal zone.

In Rhode Island, coastal marshes typically developed along tidal rivers, estuarine embayments (e.g., coves), and coastal ponds (Figure 7). Along Narragansett Bay and various coastal rivers, such as the Pettaquamscutt and Warren Rivers, coastal wetlands formed in areas of sedimentation. Sediments are transported by rivers and streams flowing seaward as well as by inflowing ocean currents. When the river meets the sea, sediments begin to settle out of suspension forming deltas and bars at the river's mouth and intertidal flats in adjacent protected areas. Sedimentation also takes place further upstream when tidal currents slow, as during slack water periods. The rate and extent of sedimentation depends on the original size and age of the estuary, present erosion rate upstream, and deposition by the river and marine tides and currents (Reid 1961). Orson and others (1987) described the formation of tidal marshes in a drowned river valley in the Pataguanset River of eastern Connecticut. The salt marsh began forming about 3,500 years ago. Halophytic (salt-tolerant) plants replaced freshwater marsh plants as salinity increased due to rising sea level (coastal submergence) and replaced upland vegetation as low-lying uplands were submerged by estuarine waters.

Coastal wetlands are common in the intertidal zone along salt and brackish "ponds" which occur behind the barrier beaches along Rhode Island's South Shore. Redfield (1972) has described development of a New England salt marsh behind a developing barrier beach. Initially, mud and silt are deposited to form tidal flats in shallow areas. As elevations exceed mean sea level, smooth cordgrass (*Spartina alterniflora*) becomes established, forming the low or regularly flooded marsh. The presence of this vegetation further slows the velocity of flooding waters, causing more sedimentation. Sediments continue to build up to a level where erosion and deposition are in relative equilibrium. The high or irregularly flooded salt marsh begins to form where the substrate builds up above the mean high water mark.

Salt marshes are still forming along the Rhode Island coast. In coastal ponds, new sediments come mainly from stormwater washover during hurricanes and other severe storms and from sea water moving through breachways, with very little input from adjacent uplands (Lee 1980; Dillon 1970; Conover 1961). These processes form extensive shoals and sand bars in coastal ponds. Coastal marshes may, in time, develop in these areas as can be witnessed inside the Charlestown Breachway (F. Golet, pers. comm.).

Wetland Hydrology

The presence of water from stream or lake flooding, surface water runoff, ground-water discharge, or tides is the driving force creating and maintaining wetlands. Hydrology determines the nature of the soils and the types of plants and animals living in wetlands. An accurate assessment of hydrology unfortunately requires extensive knowledge of the frequency and duration of flooding, water table fluctuations and ground-water relationships. This information can only be gained through intensive and long-term studies. There are, however, ways to recognized broad differences in hydrology or water regime. At certain times of the year, such as during spring floods or high tides in coastal areas, hydrology is apparent. Yet, for most of the year, such obvious evidence is lacking in many wetlands. At these times, less conspicuous signs of flooding may be observed, including (1) water marks on vegetation, (2) water-transported debris on plants or collected around their bases, and (3) water-stained leaves on the ground (Tiner 1988). These and other signs, such as wetland vegetation, help us recognize hydrologic differences between wetlands.

The Service's wetland classification system (Cowardin, et al. 1979) includes water regime modifiers to describe hydrologic characteristics. Two groups of water regimes are identified: (1) tidal and (2) nontidal. Tidal water regimes are driven by oceanic tides, while nontidal regimes are largely influenced by surface water runoff and ground-water discharge.

Tidal Wetland Hydrology

In coastal areas, ocean-driven tides are the dominant hydrologic feature of wetlands. Along the Atlantic coast, tides are semidiurnal and symmetrical with a period of 12 hours and 25 minutes. In other words, there are roughly two high tides and two low tides each day. Since the tides are largely controlled by the position of the moon relative to the sun, the highest and lowest tides (i.e., "spring tides") usually occur during full and new moons. Coastal storms can also cause extreme high and low tides. Strong winds over a prolonged period have a great impact on the

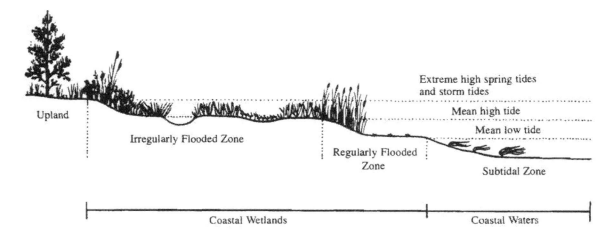

Figure 10. Hydrology of coastal wetlands showing different zones of flooding. The regularly flooded zone is flooded at least once daily by the tides, while the irregularly flooded zone is flooded less often. (Source: Tiner 1987)

normal tidal range in large coastal bays. Table 8 shows examples of varying tidal ranges along the Rhode Island coast.

In coastal wetlands, differences in hydrology (tidal flooding) create two readily identifiable zones: (1) regularly flooded zone and (2) irregularly flooded zone (Figure 10). The regularly flooded zone is alternately flooded and exposed at least once daily by the tides. It includes both the "low marsh" and intertidal mud and sand flats. Above the regularly flooded zone, the marsh is less frequently flooded by the tides (less than once a day). This irregularly flooded zone, or "high marsh", is exposed to the air for variable periods; it is usually flooded only for brief periods. Most of the high marsh is flooded during spring tides. The upper margins of the high marsh may be flooded, however, only during storm tides which are more frequent in winter. Estuarine plants have adapted to these differences in hydrology (Adams 1963; Nixon 1982) and certain plants are good indicators of different water regimes (Table 9). The tall form of smooth cordgrass (*Spartina alterniflora*) has been shown to be a reliable indicator of the landward extent of mean high tide (Kennard, *et al*. 1983).

Nontidal Wetland Hydrology

Beyond the influence of the tides, two hydrologic forces regulate water levels or soil saturation in wetlands: (1) surface water runoff and (2) ground-water discharge. Wind driven-waves across lakes cause flooding of shoreline wetlands. Surface water runs off from the land and either collects in depressional wetlands or enters rivers and lakes after snowmelt or rainfall periods and may overflow into adjacent floodplains (Figure 11). Ground water discharges into depressional wetlands when directly connected to the water table or into sloping wet-

Table 8. Tidal ranges of mean and spring tides and mean tide level at various locations in Rhode Island (U.S. Department of Commerce 1988).

Location	Mean Tide Range (ft)	Spring Tide Range (ft)	Mean Tide Level (ft)
Narragansett Bay			
at Sakonnet	3.1	3.9	1.7
at Newport	3.5	4.4	1.9
at Bristol Ferry	4.1	5.1	2.2
at Providence	4.5	5.6	2.4
at Wickford	3.8	4.7	2.0
Seekonk River at Pawtucket	4.6	5.8	2.5
Great Salt Pond at Block Island	2.6	3.2	1.4
Block Island Sound at Watch Hill Point	2.6	3.2	1.4
Pawcatuck River at Westerly	2.7	3.2	1.5

Table 9. Examples of plant indicators of tidal water regimes for Rhode Island's estuarine wetlands. These plants are generally good indicators of the regimes.

Water Regime	Indicator Plants
Regularly Flooded	Smooth Cordgrass (tall form) (*Spartina alterniflora*)
	Rockweeds (*Fucus* spp. and *Ascophyllum nodosum*)
Irregularly Flooded	Salt Hay Grass (*Spartina patens*)
	Spike Grass (*Distichlis spicata*)
	Black Grass (*Juncus gerardii*)
	Smooth Cordgrass (short form) (*S. alterniflora*)
	Switchgrass (*Panicum virgatum*)
	Common Reed (*Phragmites australis*)

lands in "seepage" areas (Figure 12). An individual wetland may exist due to surface water runoff or groundwater discharge or both. The role of hydrology in maintaining freshwater wetlands is discussed by Gosselink and Turner (1978).

Freshwater rivers and streams usually experience greatest flooding in winter and early spring, with maximum flooding usually occurring in March and April. Major flooding is frequently associated with frozen soil, snowmelt, and/or spring rains. Flooding may result from overflow of backwater streams rather than from overflow of the mainstem river (Buchholz 1981). Backwater stream levees are lower in elevation and are easily breached by rising waters. Minor drainage within the floodplain may, therefore, significantly affect flooding and drainage patterns.

Water tables fluctuate markedly during the year in many areas (Figure 13). From winter to mid-spring or early summer, the water table is at or near the surface in many wetlands. During this time, water may pond or flood the wetland surface for variable periods. In May or June, the water table may begin to drop, usually reaching its low point between late August and October. Longer days, increasing air temperatures, increasing evapotranspiration, and other factors are responsible for the consistent lowering of the water table from spring through summer.

Lowry (1984) reported on water regimes of 12 forested wetlands in southern Rhode Island: six red maple swamps and six Atlantic white cedar swamps. On average, the cedar swamps were flooded 49 percent of the growing season, while the red maple swamps were flooded only 27 percent of this time. Water levels in both types of swamps were lowest in September and usually attained their highest levels by late December. This pattern is consistent with the water table fluctuations observed by Lyford (1964) and shown in Figure 13 (year 1). Other New England studies of nontidal wetland hydrology include Anderson and others (1980), Swift and others (1984), Motts and O'Brien (1981), O'Brien (1977), Holzer (1973), and Hall and others (1972).

Standing water may be present in depressional, streamside, or lakefront wetlands for variable periods during the growing season. When flooding or ponding is brief (usually two weeks or less), the wetland is considered temporarily flooded. During the summer, the water table may drop to three feet or more below the surface in these wetlands. This situation is prevalent along floodplains. Flooding for longer periods is described by three common water regimes: (1) seasonally flooded, (2) semipermanently flooded, and (3) permanently flooded (Cowardin, et al. 1979). A seasonally flooded wetland typically has standing water visible for more than one month during the growing season, but usually by late summer, such water is absent. When not flooded, however, the water table remains within 1.5 feet of the surface for significant periods in many seasonally flooded wetlands. By contrast,

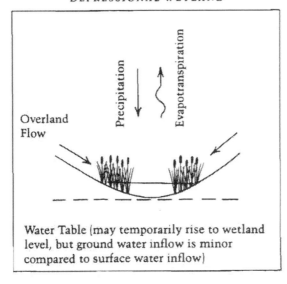

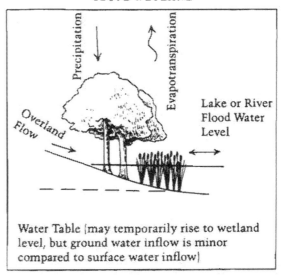

Figure 11. Hydrology of surface water wetlands. (Source: redrawn from Novitski 1982)

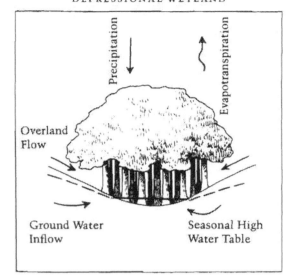
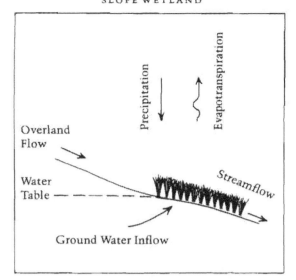

Figure 12. Hydrology of ground-water wetlands. (Source: redrawn from Novitski 1982)

a semipermanently flooded wetlands remain flooded throughout the growing season in most years. Only during dry periods does the surface of these wetlands become exposed to air. Even then, the water table lies at or very near the surface. The wettest wetlands are permanently flooded. These areas include open water bodies where depth is less than 6.6 feet, e.g., ponds and shallow portions of lakes, rivers and streams.

Some rarely flooded wetlands are almost entirely influenced by ground-water discharge or surface water runoff. Many of these wetlands occur on considerable slopes in association with springs (i.e., points of active ground-water discharge), which are commonly called "seeps". Their soils are saturated to the surface for much of the growing season and the water regime is, therefore, classified as saturated. Some seepage areas are more seasonal in nature and many "wet meadows" may be saturated only during the early part of the growing season. Other saturated wetlands include "bogs". In these situations, the soil is virtually continuously saturated.

For Rhode Island, the most common nontidal water regimes, listed by frequency of occurrence are: (1) seasonally flooded, (2) permanently flooded, (3) saturated, and (4) semipermanently flooded. Common indicator plants of nontidal water regimes are presented in Table 10. Hydrologic conditions, e.g., water table fluctuation, flooding, and soil saturation, for each of Rhode Island's

Table 10. Examples of plant indicators of nontidal water regimes for Rhode Island's wetlands.

Water Regime	Indicator Plants
Permanently Flooded	White Water Lily (*Nymphaea odorata*) Pondweeds (*Potamogeton* spp.) Water Shield (*Brasenia schreberi*) Spatterdock (*Nuphar luteum*)
Semipermanently Flooded	Buttonbush (*Cephalanthus occidentalis*) Burreeds (*Sparganium* spp.) Bayonet Rush (*Juncus militaris*) Pickerelweed (*Pontederia cordata*)
Seasonally Flooded	Broad-leaved Cattail (*Typha latifolia*) Tussock Sedge (*Carex stricta*) Turk's-cap Lily (*Lilium canadense michiganese*) Marsh Fern (*Thelypteris thelypteroides*) Arrow-leaved Tearthumb (*Polygonum sagittatum*) Highbush Blueberry (*Vaccinium corymbosum*) Swamp Rose (*Rosa palustris*) Alders (*Alnus* spp.) Common Elderberry (*Sambucus canadensis*) Atlantic White Cedar (*Chamaecyparis thyoides*)
Saturated	Pitcher Plant (*Sarracenia purpurea*) White Beak-rush (*Rhynchospora alba*) Beaked Sedge (*Carex rostrata*) Leatherleaf (*Chamaedaphne calyculata*)

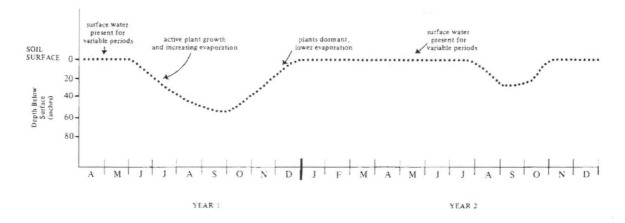

Figure 13. Water table fluctuations in a nontidal wetland (adapted from data by Lyford 1964). In general, the water table is at or near the surface through the winter and spring, drops markedly in summer, and begins to rise in the fall. As shown, the water table fluctuates seasonally and annually.

hydric soils are generally discussed in the following chapter.

For more detailed information on wetland hydrology, the reader is referred to the following sources. Some of the most current information on wetland hydrology is found in the recently published proceedings of a national symposium on this topic (Kusler and Brooks 1988).

References

Adams, D.A. 1963. Factors influencing vascular plant zonation in North Carolina salt marshes. Ecology 44(3): 445–456.

Anderson, P.H., M.W. Lefor, and W.C. Kennard. 1980. Forested wetlands in eastern Connecticut: their transition zones and delineation. Water Resources Bulletin 16(2): 248–255.

Bloom, A.L. and C.W. Ellis, Jr. 1965. Postglacial stratigraphy and morphology of Coastal Connecticut. Conn. Geol. Nat. Hist. Survey Guidebook No. 1. Hartford, CT. 10 pp.

Carter, V., M.S. Bedinger, R.P. Novitski, and W.O. Wilen. 1979. Water resources and wetlands (theme paper). *In:* Greeson, P.E., J.R. Clark and J.E. Clark (editors). Wetland Functions and Values: The State of Our Understanding. American Water Resources Association, Minneapolis, Minnesota. pp. 344–376.

Conover, R.J. 1961. A Study of Charlestown and Green Hills Ponds, Rhode Island. Ecology 42: 119–140.

Cowardin, L.W., V. Carter, F.C. Golet, and E.T. LaRoe. 1979. Classification of Wetlands and Deepwater Habitats of the United States. U.S. Fish and Wildlife Service. FWS/OBS-79/31. 103 pp.

Dansereau, P. and F. Segadas-Vianna. 1952. Ecological study of the peat bogs of eastern North America. Can. J. Bot. 30: 490–520.

Dillon, W.P. 1970. Submergence effects on a Rhode Island barrier and lagoon and interferences on migration of barriers. J. Geology 78: 94–106.

Gosselink, L.G. and R.E. Turner. 1978. The role of hydrology in freshwater wetland ecosystems. *In:* Good, R.E., D.F. Whigham, and R.L. Simpson (editors). Freshwater Wetlands. Ecological Processes and Management Potential. Academic Press, Inc., New York. pp. 63–78.

Hall, F.R., R.J. Rutherford, and G.L. Byers. 1972. The Influence of a New England Wetland on Water Quantity and Quality. University of New Hampshire, Water Resource Research Center, Durham. Res. Rept. No. 4. 51 pp.

Hicks, S.D., H.A. DeBaugh, and L.H. Hickman. 1983. Sea Level Variations for the United States: 1855–1980. U.S. Dept. of Commerce, NOAA-NOS, Rockville, MD.

Holzer, T.L. 1973. Inland wetlands and ground water in eastern Connecticut. *In:* T. Helfgott, M.W. Lefor, and W.C. Kennard (editors). Proceedings: First Wetlands Conference. University of Connecticut, Institute of Water Resources, Storrs. pp. 66–81.

Kennard, W.C., M.W. Lefor, and D.L. Civco. 1983. Analysis of Coastal Marsh Ecosystems: Effects of Tides on Vegetational Change. University of Connecticut, Institute of Water Resources, Storrs. Res. Proj. Tech. Completion Rept. B-014 CONN. 140 pp.

Kusler, J.A., and G. Brooks (editors). 1988. Proceedings of the National Wetland Symposium: Wetland Hydrology. Association of State Wetland Managers, Berne, NY. 339 pp.

Lee, V. 1980. An Elusive Compromise: Rhode Island Coastal Ponds and Their People. Coastal Resources Center, University of Rhode Island, Narragansett. Marine Tech. Rept. 73. 82 pp.

Leitch, J.A. 1981. Wetland Hydrology: State-of-the-Art and Annotated Bibliography. North Dakota State Univ., Agric. Expt. Stat. North Dakota Research Rept. No. 82. 16 pp.

Lowry, D.J. 1984. Water Regimes and Vegetation of Rhode Island Forested Wetlands. M.S. thesis, University of Rhode Island, Plant and Soil Science, Kingstown. 174 pp.

Lyford, W.H. 1964. Water Table Fluctuations in Periodically Wet Soils of Central New England. Harvard University. Harvard Forest Paper No. 8. 15 pp.

Motts, W.S. and A.L. O'Brien. 1981. Geology and Hydrology of Wetlands in Massachusetts. University of Massachusetts, Water Resources Research Center, Amherst. Publication No. 123.

Nichols, G.E. 1915. The vegetation of Connecticut. V. Plant societies along rivers and streams. Bull. Torrey Bot. Club 42: 169–217.

Nixon, S.W. 1982. The ecology of New England High Salt Marshes: A Community Profile. U.S. Fish and Wildlife Service, Washington, DC. FWS/OBS-81/55. 70 pp.

Novitzki, R.P. 1982. Hydrology of Wisconsin Wetlands. U.S. Geological Survey, Reston, VA. Information Circular 40. 22 pp.

Novitski, R.P. 1989. Wetland Hydrology. Chapter 5. *In:* S.K. Majumdar, R.P. Brooks, F.J. Brenner, and R.W. Tiner, Jr. (editors). Wetlands Ecology and Conservation: Emphasis in Pennsylvania. The Pennsylvania Academy of Sciences. pp. 47–65.

O'Brien, A.L. 1977. Hydrology of two small wetland basins in eastern Massachusetts. Water Resources Bulletin 13: 325–340.

Orson, R.A., R.S. Warren, and W.A. Niering. 1987. Development of a tidal marsh in a New England river valley. Estuaries 10(1): 20–27.

Quinn, A.W. 1973. Rhode Island Geology for the Non-Geologist. RI Dept. of Natural Resources, Providence. 63 pp.

Redfield, A.C. 1972. Development of a New England salt marsh. Ecol. Monogr. 42: 201–237.

Reid, G.K. 1961. Ecology of Inland Waters and Estuaries. Van Nostrand Reinhold Co., New York. 375 pp.

Shepps, V.C. 1978. Pennsylvania and the Ice Age. Pennsylvania Dept. of Environmental Resources, Bureau of Topographic and Geologic Survey, Harrisburg. Educ. Series No. 6. 33 pp.

Swift, B.L., J.S. Larson, and R.M. DeGraaf. 1984. Relationship of breeding bird density and diversity to habitat variables in forested wetlands. Wilson Bulletin 96: 48–59.

Tiner, R.W., Jr. 1988. Field Guide to Nontidal Wetland Identification. Maryland Dept. of Natural Resources, Annapolis and U.S. Fish and Wildlife Service, Newton Corner, MA. Cooperative publication. 283 pp. + plates.

Titus, J.G. and S.R. Seidel. 1986. Overview of the effects of changing the atmosphere. *In:* J.G. Titus (editor). Effects of Changes in Stratospheric Ozone and Global Climate. Volume 1: Overview. U.S. Environmental Protection Agency, Washington, DC and The United Nations Environment Programme. Cooperative publication. pp. 3–19.

U.S. Department of Commerce. 1988. Tide Tables 1989. High and Low Water Predictions. NOAA, National Ocean Survey. 289 pp.

Wolfe, P.E. 1977. The Geology and Landscapes of New Jersey. Crane, Russak and Co., Inc. 351 pp.

CHAPTER 5.

Hydric Soils of Rhode Island

Introduction

The predominance of undrained hydric soil is a key attribute for identifying wetlands (Cowardin, *et al.* 1979). Hydric soils naturally develop in wet depressions, on floodplains, on seepage slopes, and along the margins of coastal and inland waters. Knowledge of hydric soils is particularly useful in distinguishing the drier wetlands from uplands, where the more typical wetland plants are less common or absent. This chapter focuses on the characteristics, distribution and extent of Rhode Island's hydric soils. Tiner and Veneman (1987) describe characteristics and field recognition of New England's hydric soils.

Definition of Hydric Soil

Hydric soils have been defined by the U.S.D.A. Soil Conservation Service (1987) as follows: "A hydric soil is a soil that is saturated, flooded or ponded long enough during the growing season to develop anaerobic conditions in the upper part." This definition includes soils that are saturated with water at or near the soil surface and virtually lacking free oxygen for a significant period of the growing season and soils that are ponded or frequently flooded for long periods during the growing season. Table 11 lists criteria for hydric soils.

Soils that were formerly wet, but are now completely drained, are not considered hydric soils or wetlands, according to the Service's wetland classification system (Cowardin, *et al.* 1979). These soils must be checked in the field to verify that drainage measures will remain functional under normal or design conditions. Where failure of a drainage system results, such soils can revert to hydric conditions. This condition must be determined on a site-specific basis. Soils that were not naturally wet, but are now subject to periodic flooding or soil saturation for specific management purposes (e.g., waterfowl impoundments) or flooded by accident (e.g., highway-created impoundments) are considered hydric soils (see criteria 3 and 4 in Table 11). Hydrophytes are usually present in these created wetlands. Better-drained soils that are frequently flooded for short intervals (usually less than one week) during the growing season, are not considered hydric soils.

Major Categories of Hydric Soils

Hydric soils are separated into two major categories on the basis of soil composition: (1) organic soils (histosols) and (2) mineral soils. In general, soils having 20 percent or more organic material by weight in the upper 16 inches are considered organic soils, while soils with less organic content are mineral soils. For a technical definition, the reader is referred to *Soil Taxonomy* (Soil Survey Staff 1975).

Accumulation of organic matter results from prolonged anaerobic soil conditions associated with long periods of flooding and/or soil saturation during the growing season. These saturated conditions impede aerobic decomposition (or oxidation) of the bulk organic materials, such as leaves, stems and roots, and encourage their accumulation as peat or muck over time. Consequently, most or-

Table 11. Criteria for hydric soils (U.S.D.A. Soil Conservation Service 1987).

Criteria for Hydric Soils

1. All Histosols (Organic soils) except Folists, or

2. Soils (Mineral soils) in Aquic suborders, Aquic subgroups, Allbolls suborder, Salorthids great groups, or Pell great groups of Vertisols that are:

    a. somewhat poorly drained and have water table less than 0.5 feet from the surface for a significant period (usually a week or more) during the growing season, or

    b. poorly drained or very poorly drained and have either:

       (1) water table at less than 1.0 feet from the surface for a significant period (usually a week or more) during the growing season if permeability is equal to or greater than 6.0 inches/hour in all layers within 20 inches, or

       (2) water table at less than 1.5 feet from the surface for a significant period (usually a week or more) during the growing season if permeability is less than 6.0 inches/hour in any layer within 20 inches, or

3. Soils (Mineral soils) that are ponded for long duration (more than 7 days) or very long duration (more than 1 month) during the growing season, or

4. Soils (Mineral soils) that are frequently flooded for long duration (more than 7 days) or very long duration (more than 1 month) during the growing season.

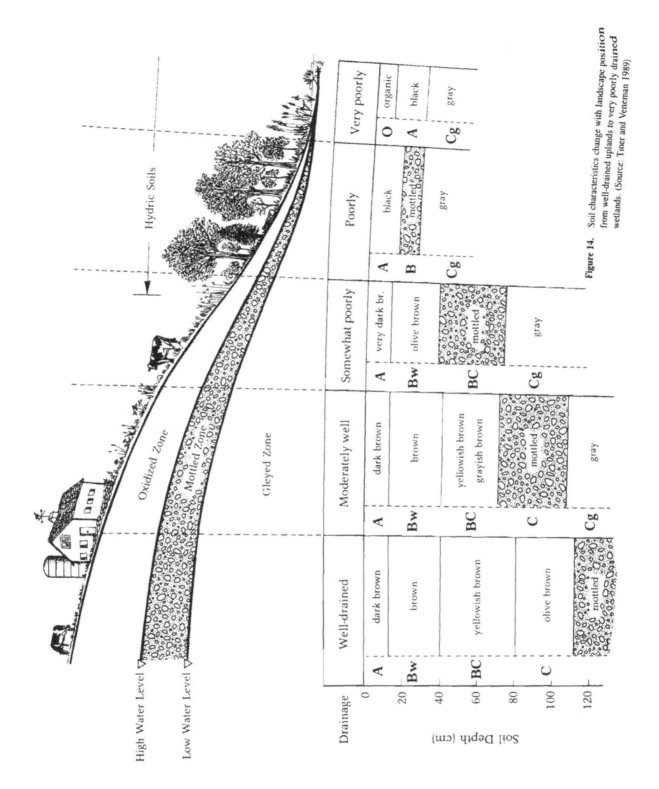

Figure 14. Soil characteristics change with landscape position from well-drained uplands to very poorly drained wetlands. (Source: Tiner and Veneman 1989)

tivity of certain soil microorganisms. These microorganisms reduce iron when the soil environment is anaerobic, that is, when virtually no free oxygen is present, and when the soil contains organic matter. If the soil conditions are such that free oxygen is present, organic matter is absent, or temperatures are too low (below 41°F or 5°C) to sustain microbial activity, gleization will not proceed and mottles will not form, even though the soil may be saturated for prolonged periods of time (Diers and Anderson 1984).

In Rhode Island, one well-drained soil may appear to be a hydric soil due to dark surface and subsurface layers. The Newport series (Typic Fragiochrepts) may have a black to dark gray or dark brown surface layer with a dark gray to dark grayish brown or olive gray subsoil (to about 19 inches deep). These dark grayish colors are *not* associated with a reducing (anaerobic) environment, but result from the soil parent materials of dark sandstone, conglomerate, argillite, and phyllite.

National List of Hydric Soils

To help the Service clarify its wetland definition, the U.S.D.A. Soil Conservation Service (SCS) agreed to develop a list of hydric soils. Work on the list began in the late 1970's and the list underwent a few revisions prior to its most recent printing in 1987. The national hydric soils list is reviewed annually and updated and republished as needed. This list summarizes in tabular form the characteristics of each designated hydric soil. State lists of hydric soils have been prepared from the national list and are available from the U.S. Fish and Wildlife Service's National Wetlands Inventory Project.

Rhode Island's Hydric Soils

In Rhode Island, 14 soil series have been identified as hydric soils. These series are very poorly drained or poorly drained soils. Table 13 lists these soils along with selected characteristics. Examples of Rhode Island's hydric soils and non-hydric soils are shown in Plates 1–6.

County Acreage of Hydric Soils

Recent SCS soil mapping in Rhode Island identified 112,690 acres of "potential" hydric soils. This represents about 17 percent of the state's land surface area. In Rhode Island, somewhat poorly drained soils were not separated from the poorly drained soils in soil mapping, therefore, the total acreage of "potential" hydric soils is actually higher than the true acreage of hydric soils associated with wetlands in the state (E. Stuart, pers. comm.). Table 14 outlines acreages of potential hydric soils for each county based on SCS soil mapping (Rector 1981).

Hydric Soil Descriptions

This subsection briefly describes key features of each hydric soil associated with Rhode Island's wetlands. This information was obtained from Rector (1981); for more detailed descriptions, please contact the Soil Conservation Service in West Warwick. (*Note:* The Wareham series is not included in the descriptions because it is of minor importance and was not mapped as a soil map unit during SCS's soil survey; its characteristics are, however, presented in Table 13, since it does occur in the state.)

Adrian Series

The Adrian series consists of very poorly drained organic soils (mucks). They are usually found in depressions and drainageways of glacial till uplands and outwash plains. This soil is characterized by a layer of shallow, black muck about 20 inches thick (range: 16 to 50 inches). The water table is at or near the surface for most of the year, usually from November through May. Flooding occurs in a few areas. Adrian muck is found in all counties, with nearly half of the state's acreage in Washington County.

Carlisle Series

Carlisle soils are very poorly drained organic soils (mucks) associated with depressions in glacial till uplands and outwash plains. These soils are represented by a deep, black and dark reddish brown muck more than 51 inches thick. The water table is at or near the surface for most of the year, usually from September into June. Flooding or ponding of surface water may occur for variable periods. Carlisle muck occurs in every county, but is most common in Washington County, which possesses over 60 percent of the state's acreage of this soil.

Ipswich Series

The Ipswich series is represented by very poorly drained organic soils (peats) associated with estuarine tidal marshes. It is characterized by a very dark brown surface peat about 11 inches thick, overlying a very dark grayish brown subsurface mucky peat about 60 inches thick. The organic layer is more than 51 inches thick. The soil is saturated with salt water and subjected to frequent tidal flooding. Although Ipswich peat occurs in Bristol, Kent, and Washington counties, over 90 percent of the state's acreage of this soil is present in Bristol County.

Table 13. Hydric soils of Rhode Island, including soil taxonomic names (series and soil subgroups), drainage class (VP—very poorly drained, P—poorly drained, SP—somewhat poorly drained), depth (below the soil surface), frequency and applicable period of high water table (+ denotes water level above the soil surface; +V indicates flooding at variable depths) or flooding, and hydric criteria number (see Table 11). An asterisk (*) indicates that the series may include non-hydric members. *Note:* Some New England soils classified as poorly drained actually include somewhat poorly drained members which may be non-hydric. (Sources: Modified from Tiner and Veneman 1987; Rector 1981).

Series	Soil Subgroup	Drainage Class	High Water Table		Flooding			Hydric Criterion Number
			Depth (ft)	Period	Frequency	Duration	Period	
Adrian	Terric Medisaprists	VP	+1.0–1.0	Nov–May	None			1
Adrian, flooded	Terric Medisaprists	VP	+1.0–0.5	Nov–May	Frequent	Long	Nov–May	1,4
Carlisle	Typic Medisaprists	VP	+0.5–1.0	Sep–Jun	None			1
Carlisle, flooded	Typic Medisaprists	VP	+0.5–1.0	Sep–Jun	Frequent	Long	Nov–May	1,4
Ipswich	Typic Sulfihemists	VP	+1.0–0.0	Jan–Dec	Frequent	V. Brief	Jan–Dec	1
*Leicester	Aeric Haplaquepts	P	0.0–1.5	Nov–Mar	None			2A;2B2
Mansfield	Typic Humaquepts	VP	+1.0–0.5	Nov–Jul	None			2B2
Matunuck	Typic Sulfaquents	VP	+1.0–0.0	Jan–Dec	Frequent	V. Brief	Jan–Dec	2B1
*Raypol	Aeric Haplaquepts	P	0.0–1.0	Nov–May	None			2A;2B2
*Ridgebury	Aeric Fragiaquepts	SP, P	0.0–1.5	Nov–May	None			2A;2B2
*Rumney	Aeric Fluvaquents	P	+V–1.5	Nov–Jun	Frequent	Brief	Oct–May	2A;2B2
Scarboro	Histic Humaquepts	VP	+1.0–1.0	Nov–Jul	Rare			2B1
*Stissing	Aeric Fragiaquepts	P	0.0–1.5	Oct–May	None			2B2
*Walpole	Aeric Haplaquents	P	0.0–1.0	Nov–Apr	None			2A;2B2
*Wareham	Humaqueptic Psammaquents	SP, P	0.0–1.5	Sep–Jun	None			2A;2B2
*Wareham, flooded	Humaqueptic Psammquents	SP, P	+V–1.5	Sep–Jun	Occasional	Brief	Mar–May	2A;2B2;4
Whitman	Typic Humaquepts	VP	+1.0–0.5	Sep–Jun	None			2B2

Table 14. County summaries of hydric soils acreage in Rhode Island (taken from Rector 1981). *Note:* Soil map units that include somewhat poorly drained soils that are usually non-hydric.

Soil Map Unit Name	Bristol County	Kent County	Newport County	Providence County	Washington County	State Totals	% of State Covered
Adrian muck	45	2,540	530	2,505	5,180	10,800	1.6
Carlisle muck	5	1,790	850	3,070	9,255	14,970	2.2
Ipswich peat	525	30	—	—	20	575	0.1
Mansfield mucky silt loam	65	—	1,940	5	200	2,210	0.3
Mansfield very stony mucky silt loam	15	—	1,630	—	105	1,750	0.3
Matunuck mucky peat	390	175	1,070	115	1,805	3,555	0.5
Raypol silt loam*	15	285	15	55	2,090	2,460	0.4
Ridgebury fine sandy loam*	105	80	275	1,450	245	2,155	0.3
Ridgebury, Whitman and Leicester extremely stony fine sandy loams*	390	6,955	805	26,040	11,230	45,420	6.7
	15	415	25	750	795	2,000	0.3
Rumney fine sandy loam*	15	415	25	750	795	2,000	0.3
Scarboro mucky sandy loam	260	1,330	585	1,720	4,075	7,970	1.2
Stissing silt loam*	395	40	6,885	35	1,485	8,840	1.3
Stissing very stony silt loam*	245	—	730	40	480	1,495	0.2
Walpole sandy loam*	475	1,805	90	3,100	3,020	8,490	3.3
Totals	2,945	15,445	15,430	38,885	39,985	112,690	16.6

Leicester Series

The Leicester series consists of poorly drained mineral soils (extremely stony fine sandy loams). These soils are found in depressions and drainageways in glacial till uplands. They are characterized by a very dark grayish brown fine sandy loam surface layer about 8 inches thick and a light brownish gray and light yellowish brown mottled fine sandy loam subsurface layer about 18 inches thick over a gray, mottled gravelly sandy loam to a depth of 60 inches or more. Stones and boulders cover 10–35 percent of the mapped area of this soil. The water table is within 1.5 feet of the surface from November into March and ponding or flooding does not occur, yet the soils are saturated for long periods. Leicester soils are mapped as a complex with Ridgebury and Whitman soils; it has been mapped in each county, but is most abundant in Providence County.

Mansfield Series

Mansfield soils are very poorly drained mineral soils (mucky silt loams or very stony mucky silt loams). They are associated with depressions and small drainageways of drumlins in the southeastern part of the state. These mineral soils are characterized by an 8-inch thick black mucky silt loam surface layer overlying a dark gray and olive gray silt loam subsurface layer to a depth of 60 inches or more. The water table is within 0.5 feet of the surface from November into July. Mansfield soils are most abundant in Newport County, which has 90 percent of the state's acreage of this soil, but they also occur to a limited extent in Washington, Bristol, and Providence Counties.

Matunuck Series

Although called mucky peat, the Matunuck series consists of very poorly drained mineral soils having shallow organic surface layers. They are found in estuarine tidal marshes. These soils are characterized by a shallow surface layer of very dark gray mucky peat about one foot thick overlying a gray sand subsurface layer to a depth of 60 inches or more. The water table is at the surface and tidal flooding with salt water is frequent throughout the year. Matunuck mucky peat occurs in every county, with about 80 percent of the state's acreage of this soil in Washington and Newport Counties.

Raypol Series

Raypol soils are poorly drained mineral soils (silt loams). They are found in depressions or terraces and outwash plains. Raypol soils are represented by a very dark grayish brown silt loam shallow surface layer about 4 inches thick and an 18-inch subsurface layer of light olive brown mottled silt loam, overlying a grayish brown and yellowish brown, mottled gravelly sand to a depth of 60 inches or more. The water table is within one foot of the surface from November into May and flooding or ponding does not take place. Raypol soils are found in all counties, but are mostly in Washington County which possesses about 85 percent of the state's acreage of this soil.

Ridgebury Series

The Ridgebury series contains poorly drained mineral soils with a 10–35% areal coverage of stones and boulders (extremely stony fine sandy loams). These soils are associated with depressions and drainageways in glacial till uplands. Ridgebury soils are characterized by a 4-inch thick black fine sandy loam surface layer, a 16-inch grayish brown fine sandy loam subsoil layer that is mottled in the lower portion, and a yellowish brown substratum of mottled gravelly fine sandy loam to a depth of 60 inches or more. A perched water table is within 1.5 feet from November into May. These soils were mapped as part of a soil complex with Whitman and Leicester. This complex was mapped in all counties, with more than half of the state's acreage of this type mapped in Providence County.

Rumney Series

Rumney soils are poorly drained mineral soils (fine sandy loams). They are found in recent alluvium on floodplains. They are characterized by a very dark grayish brown fine sandy loam surface layer about 5 inches thick, a dark grayish brown, mottled fine sandy loam about 17 inches thick, and a gray and dark grayish brown sand substratum 60 inches or greater in thickness. The high water table is at about 0.5 feet from November into June. The soil is also subject to frequent brief floods. Rumney soils are found in every county but are most abundant in Washington and Providence Counties which contain about 77 percent of the state's acreage of this soil.

Scarboro Series

Scarboro soils are very poorly drained mineral soils, with a high content of organic matter (mucky sandy loams). They are found in drainageways and depressions on terraces and outwash plains. These soils are characterized by a 6-inch thick very dark grayish brown mucky sandy loam surface layer overlying a deep substratum composed of two layers: the upper layer—a gray, mottled loamy sand and the lower layer—a light brownish gray, mottled coarse sand. In some areas, the soil may have an organic layer (mucky peat or muck) up to 16 inches thick. The water table is within a foot of the surface from

November into July and ponding with up to one foot of surface water may occur. Flooding takes place in a few areas. Scarboro soils are mapped in every county, but more than half of the state's acreage of this soil is found in Washington County.

Stissing Series

The Stissing series consists of poorly drained mineral soils (silt loams or very stony silt loams, depending upon the extent of stones and boulders present). These soils occur in the southeastern part of the state on glacial upland hills and drumlins. They are characterized by a very dark gray silt loam surface layer about 8 inches thick overlying a 7-inch thick dark grayish brown mottled silt loam subsoil and a dark gray, mottled silt loam substratum extending to a depth of 60 inches or more. A perched water table is within 1.5 feet from October into May. Stissing soils have been mapped in each county, but nearly 75 percent of the state's acreage of this soil is found in Newport County.

Walpole Series

Walpole soils are poorly drained mineral soils (sandy loams) associated with depressions and small drainageways on terraces and outwash plains. They are characterized by a 7-inch thick very dark brown sandy loam surface layer, a 12-inch thick light brownish gray, mottled sandy loam subsoil, and a dark yellowish brown and grayish brown, mottled gravelly sand substratum extending to a depth of 60 inches or more. The water table lies within one foot, often within 6 inches, of the surface from November into April. Walpole sandy loams occur in all counties, but are most prevalent in Providence and Washington Counties, which comprise nearly 75 percent of the state's acreage of this soil.

Whitman Series

The Whitman series consists of very poorly drained mineral soils with a 10–35 percent areal coverage by stones and boulders (extremely stony fine sandy loams). These soils are found in drainageways and depressions in glacial till uplands. They are characterized by a 10-inch thick surface layer of black fine sandy loam overlying a gray gravelly fine sandy loam substratum that is mottled for more than 18 inches. The water table is within 0.5 feet of the surface from September into June and surface water may pond to a depth of one foot. Whitman soils are mapped as a complex with Ridgebury and Leicester soils and are found in every county, with over half of the state's acreage of this complex mapped in Providence County.

References

Aandahl, A.R., S.W. Buol, D.E. Hill, and H.H. Bailey (editors). 1974. Histosols: Their Characteristics, Classification, and Use. Soil Sci. Soc. Am. Special Pub. Series No. 6. 136 pp.

Bouma, J. 1983. Hydrology and soil genesis of soils with aquic moisture regimes. In: L.P. Wilding, N.E. Smeck, and G.F. Hall (editors). Pedogenesis and Soil Taxonomy I. Concepts and Interactions. Elsevier Science Publishers, B.V. Amsterdam. pp. 253–281.

Cowardin, L.M., V. Carter, F.C. Golet and E.T. LaRoe. 1979. Classification of Wetlands and Deepwater Habitats of the United States. U.S. Fish and Wildlife Service. FWS/OBS-79/31. 103 pp.

Diers, R. and J.L. Anderson. 1984. Part I. Development of Soil Mottling. Soil Survey Horizons (Winter): 9–12.

Parker, W.B., S. Faulkner, B. Gambrell, and W.H. Patrick, Jr. 1984. Soil wetness and aeration in relation to plant adaptation for selected hydric soils of the Mississippi and Pearl River Deltas. In: Proc. of Workshop on Characterization, Classification, and Utilization of Wetland Soils (March 26–April 24). Internat. Rice Res. Inst., Los Banos, Laguna, Philippines.

Ponnamperuma, F.N. 1972. The chemistry of submerged soils. Advances in Agronomy 24: 29–96.

Rector, D.D. 1981. Soil Survey of Rhode Island. U.S.D.A. Soil Conservation Service, West Warwick, RI. 200 pp. + maps.

Soil Survey Staff. 1951. Soil Survey Manual. U.S. Department of Agriculture, Soil Conservation Service, Washington, DC. Agriculture Handbook No. 18. 502 pp.

Soil Survey Staff. 1975. Soil Taxonomy. U.S. Department of Agriculture, Soil Conservation Service, Washington, DC. Agriculture Handbook No. 436. 754 pp.

Tiner, R.W., Jr. and P.L.M. Veneman. 1987. Hydric Soils of New England. University of Massachusetts Cooperative Extension, Amherst. Bulletin C-183. 27 pp. (Note: Reprinted with minor updating in 1989.)

U.S.D.A. Soil Conservation Service. 1987. Hydric Soils of the United States. In cooperation with the National Technical Committee for Hydric Soils. Washington, DC.

Veneman, P.L.M., M.J. Vepraskas, and J. Bouma. 1976. The physical significance of soil mottling in a Wisconsin toposequence. Geoderma 15: 103–118.

CHAPTER 6.
Vegetation and Plant Communities of Rhode Island's Wetlands

Introduction

Rhode Island's wetlands are largely colonized by plants adapted to existing hydrologic, water chemistry, and soil conditions, although some wetlands (e.g., tidal mud flats) are devoid of macrophytic plants. Most wetland definitions have relied heavily on dominant vegetation for identification and classification purposes. The presence of "hydrophytes" or wetland plants is one of the three key attributes of the Service's wetland definition (Cowardin, et al. 1979). Vegetation is usually the most conspicuous feature of wetlands and one that may be often readily identified in the field. In this chapter, after discussing the concept of "hydrophyte," attention will focus on the major plant communities of Rhode Island's wetlands. The Appendix contains a list of plant species that occur in Rhode Island's wetlands.

Hydrophyte Definition and Concept

Wetland plants are technically referred to as "hydrophytes." The Service defines a "hydrophyte" as "any plant growing in water or on a substrate that is at least periodically deficient in oxygen as a result of excessive water content" (Cowardin, et al. 1979). Thus, hydrophytes are not restricted to true aquatic plants growing in water (e.g., ponds, lakes, rivers, and estuaries), but also include plants morphologically and/or physiologically adapted to periodic flooding or saturated soil conditions typical of marshes, swamps, bogs, and bottomland forests. The concept of hydrophyte applies to individual plants and not to species of plants, although certain species may be represented entirely by hydrophytes, such as smooth cordgrass (*Spartina alterniflora*) and broad-leaved cattail (*Typha latifolia*). For example, certain individuals of white pine (*Pinus strobus*) can be considered hydrophytes since they grow in undrained hydric soils (Tiner 1988a). Wet ecotypes of many plant species undoubtedly exist. All plants growing in wetlands have adapted in one way or another for life in periodically flooded or saturated, anaerobic soils. Consequently, these individuals are considered hydrophytes.

The Service has prepared a comprehensive list of plant species that are found in the Nation's wetlands to help clarify its wetland definition (Reed 1988). A list of plant species that occur in Rhode Island's wetlands has been extracted from the national list and is included in the Appendix of this report. This list contains 1,335 species of plants that may occur in Rhode Island's wetlands, including 74 species of aquatics, 52 species of ferns and fern allies, 121 species of grasses, 150 species of sedges, 21 species of rushes, 664 species of forbs (other herbaceous plants), 131 species of shrubs, 93 species of trees, and 28 species of vines. The Service recognizes four types of plants that occur in wetlands: (1) obligate wetland (OBL), (2) facultative wetland (FACW), (3) facultative (FAC), and (4) facultative upland (FACU). Obligate hydrophytes are those plants which nearly always (more than 99% of the time) occur in wetlands under natural conditions. The facultative types can be found in both wetlands and uplands to varying degrees. Facultative wetland plants usually occur in wetlands (from 67 percent to 99 percent of the time), while purely facultative plants show no affinity to wetlands or uplands (equally likely to occur in both habitats) and are found in wetlands with a frequency of occurrence between 34–66 percent. By contrast, facultative upland plants usually occur in uplands, but are present in wetlands between 1–33 percent of the time. When present, they are often in drier wetlands where they may dominate or at higher elevations (e.g., hummocks) in wetter areas. Table 15 shows the number of plant species in each wetland indicator status category. Examples of the four major types of wetland plants for Rhode Island are presented in Table 16. Field guides for identifying wetland plants are available (Magee 1981; Tiner 1987, 1988b).

Table 15. Wetland indicator status of various life forms of Rhode Island's wetland plants. Number of species in each category is listed. Plants not assigned an indicator status are listed under NI (No Indicator). *Note:* Categories FACW, FAC, and FACU include those species designated with a + and – on the list in the Appendix. (P. Reed, pers. comm.)

Life Form	OBL	FACW	FAC	FACU	NI
Aquatics	74	—	—	—	—
Ferns and Allies	8	19	13	12	—
Grasses	23	32	22	43	1
Sedges	96	33	9	12	1
Rushes	8	7	4	2	—
Forbs	235	127	107	183	12
Shrubs	22	29	35	39	6
Trees	5	24	23	38	3
Vines	—	2	10	14	2
	471	273	223	343	25

Table 16. Examples of four wetland plant types occurring in Rhode Island.

Hydrophyte Type	Plant Common Name	Scientific Name
Obligate	Royal Fern	*Osmunda regalis*
	White Water Lily	*Nymphaea odorata*
	Smooth Cordgrass	*Spartina alterniflora*
	Bluejoint	*Calamagrostis canadensis*
	Tussock Sedge	*Carex stricta*
	Three-way Sedge	*Dulichium arundinaceum*
	Broad-leaved Cattail	*Typha latifolia*
	Water Willow	*Decodon verticillatus*
	Leatherleaf	*Chamaedaphne calyculata*
	Big Cranberry	*Vaccinium macrocarpon*
	Buttonbush	*Cephalanthus occidentalis*
	Atlantic White Cedar	*Chamaecyparis thyoides*
Facultative Wetland	Cinnamon Fern	*Osmunda cinnamomea*
	Salt Hay Grass	*Spartina patens*
	Common Reed	*Phragmites australis*
	Boneset	*Eupatorium perfoliatum*
	Reed Canary Grass	*Phalaris arundinaceum*
	Speckled Alder	*Alnus rugosa*
	Highbush Blueberry	*Vaccinium corymbosum*
	Steeplebush	*Spiraea tomentosa*
	Green Ash	*Fraxinus pennsylvanica*
	American Elm	*Ulmus americana*
	Rose Bay	*Rhododendron maximum*
Facultative	Foxtail Grass	*Setaria geniculata*
	Wrinkled Goldenrod	*Solidago rugosa*
	Purple Joe-Pye-weed	*Eupatoriadelphus purpureus*
	Poison Ivy	*Toxicodendron radicans*
	Sweet Pepperbush	*Clethra alnifolia*
	Sheep Laurel	*Kalmia angustifolia*
	Oblong-leaf Shadbush	*Amelanchier canadensis*
	Red Maple	*Acer rubrum*
	Yellow Birch	*Betula alleghaniensis*
	Ironwood	*Carpinus caroliniana*
Facultative Upland	Ground-pine	*Lycopodium obscurum*
	Bracken Fern	*Pteridium aquilinium*
	Partridgeberry	*Mitchella repens*
	Black Huckleberry	*Gaylussacia baccata*
	American Holly	*Ilex opaca*
	White Ash	*Fraxinus americana*
	White Pine	*Pinus strobus*
	Hemlock	*Tsuga canadensis*

Wetland Plant Communities

Many factors influence wetland vegetation and community structure, including climate, hydrology, water chemistry, and human activities. Penfound (1952) identified the most important physical factors as: (1) location of the water table, (2) fluctuation of water levels, (3) soil type, (4) acidity, and (5) salinity. He also recognized the role of biotic factors, i.e., plant competition, animal actions, and human activities. Many construction projects alter the hydrology of wetlands through channelization and drainage or by changing surface water runoff patterns, especially in urban areas. These activities often have a profound effect on plant composition. This is particularly evident in coastal marshes where mosquito ditching has increased the abundance of high-tide bush (*Iva frutescens*), especially on spoil mounds adjacent to ditches. Repeated timber cutting and severe fires may also have profound effects on wetland communities. Many former Atlantic white cedar swamps in the Northeast are now red maple swamps because most of the cedar has been harvested.

Rhode Island's wetlands fall within five ecological systems inventoried by the NWI: Marine, Estuarine, Riverine, Lacustrine and Palustrine. In coastal areas, the estuarine marshes, which include salt marshes and tidal mudflats, are most abundant, with marine wetlands generally limited to rocky shores and intertidal beaches along Block Island and Rhode Island Sounds. Overall, however, palustrine wetlands predominate, representing about 87 percent of the state's wetlands, whereas estuarine wetlands represent only 11 percent. Palustrine wetlands include the overwhelming majority of freshwater marshes, swamps, and ponds. Wetlands associated with the riverine and lacustrine systems are largely restricted to nonpersistent emergent wetlands, aquatic beds, and nonvegetated flats. The following sections address major wetland types in each ecological system. Descriptions are based on field observations and a review of scientific literature.

Marine Wetlands

The Marine System includes the open ocean overlying the continental shelf and the associated high-energy coastline. Deepwater habitats predominate in this system, with wetlands generally limited to sandy intertidal beaches, cobble-gravel shores, and rocky shores along Block Island and Rhode Island Sounds and sand bars at the mouths of coastal inlets (Figure 15). Vegetation is sparse and scattered along the upper zones of beaches. Vascular plants, such as sea rocket (*Cakile edentula*), beach grass (*Ammophila breviligulata*), beach orach (*Atriplex arenaria*), and beach pea (*Lathyrus japonicus*) occur in these areas. Rocky shores are often colonized by macroalgae, mainly rockweeds (*Fucus* spp. and *Ascophyllum nodosum*).

Estuarine Wetlands

The Estuarine System consists of tidal salt and brackish waters and contiguous wetlands where ocean water is at least occasionally diluted by freshwater runoff from the land. It extends upstream in coastal rivers to fresh water where no measurable ocean-derived salts (less than 0.5 parts per thousand) can be detected. A variety of wetland types develop in estuaries largely because of differences in salinity and duration and frequency of flooding. Major estuarine wetland types include: (1) intertidal flats, (2) emergent wetlands, (3) scrub-shrub wetlands, and (4) aquatic beds.

Estuarine Intertidal Flats

Intertidal flats of mud and/or sand are extremely common in estuaries, particularly between salt marshes and coastal waters (Figure 16). Small gravel flats are present in isolated areas of Narragansett Bay (F. Golet, pers. comm.). Estuarine tidal flats are typically flooded by tides and exposed to air twice daily. These flats are generally devoid of macrophytes, although smooth cordgrass (*Spartina alterniflora*) may occur in isolated clumps on mud flats. Microscopic plants, especially diatoms, euglenoids, dinoflagellates and blue green algae, are often extremely abundant, yet inconspicuous (Whitlatch 1982).

Estuarine Emergent Wetlands

Differences in salinity and tidal flooding within estuaries have a profound and visible effect on the distribution of emergent vegetation. Plant composition changes markedly from the more saline portions to the brackish areas further inland. Even within areas of similar salinity, vegetation differs largely due to the frequency and duration of tidal flooding and, locally, due to freshwater runoff.

Salt marshes are the typical estuarine emergent wetlands in Rhode Island. They have formed on the intertidal shores of salt and brackish tidal waters in three major types of areas: (1) salt ponds (e.g., Winnapaug, Ninigret, and Potter Ponds), (2) coastal rivers (e.g., Pettaquamscutt, Barrington, and Warren Rivers), and (3) coves, harbors and other coastal embayments (e.g., Colonel Willie Cove, Coggeshall Cove, Wickford Harbor, Dutch Harbor, Mill Cut, and Mount Hope Bay). Plates 7–8 show examples of Rhode Island's salt marshes.

Based on differences in tidal flooding, two general vegetative zones are recognized within salt marshes: (1) regularly flooded low marsh and (2) irregularly flooded high marsh (Figure 17). The vegetation of each zone is different due to flooding frequency and duration, among other factors. The low marsh is flooded at least once daily by the tides. Above this level is the high marsh which is flooded less often than daily.

A single plant—the tall form (approximately 3–6 feet high) of smooth cordgrass (*Spartina alterniflora*)—dominates the low marsh from approximately mean sea level to the mean high water mark. This zone is generally limited to creekbanks and upper borders of tidal flats. A recent study in Connecticut found that the tall form of smooth cordgrass was an accurate indicator of the landward extent of mean high tide (Kennard, *et al.* 1983).

The vegetation of the high marsh often forms a complex mosaic rather than a distinct zone. Plant diversity increases with several species being abundant, including a short form of smooth cordgrass, salt hay grass (*Spartina patens*), spike grass (*Distichlis spicata*), glassworts (*Salicornia* spp.), marsh orach (*Atriplex patula*), sea lavender (*Limonium nashii*), salt marsh aster (*Aster tenuifolius*), and black grass (*Juncus gerardii*). Pools and tidal creeks

Plates 1-6. Examples of four hydric soils and two nonhydric soils. (1) Carlisle muck, (2) Scarboro mucky sandy loam (3) Wareham loamy sand, (4) Ridgebury fine sandy loam (hydric), (5) Ridgebury fine sandy loam (nonhydric), and (6) Sudbury sandy loam. Compare Plates 4 and 5 which show hydric and nonhydric members of the Ridgebury series. Note the position of the gray subsoils in each - immediately below the surface layer in the hydric soil and below 20 inches in the nonhydric soil. The nonhydric Sudbury soil is clearly much brighter colored than the dull-colored hydric soils.

Plate 7. Salt marsh (estuarine intertidal emergent wetland) in Tiverton. Note the developing marsh (clumps of smooth cordgrass) on the tidal flat.

Plate 8. Salt marsh adjacent to Winnapaug Pond, Westerly. Note the regularly flooded low marsh along the creek and mixed vegetation pattern of the irregularly flooded high marsh.

Plate 9. Freshwater wetlands in Frying Pan Pond (Wood River), Richmond. Note the distinctive plant zonation: water-hearts in shallow permanently flooded area, pickerelweed and bayonet rush in the semipermanently flooded zone, and sedges, grasses, and shrubs in the seasonally flooded area.

Plate 10. Seasonally flooded marsh (palustrine emergent wetland) along the Chipuxet River, West Kingston.

Plate 11. Seasonally flooded marsh on Block Island. Note the diverse plant community.

Plate 12. Grazed wet meadow in Tiverton. Sweet flag and soft rush dominate this saturated palustrine wetland.

Plate 13. Diamond Bog, Richmond in autumn. Leatherleaf, a broad-leaved evergreen shrub, dominates this saturated scrub-shrub wetland.

Plate 14. Newton Marsh, Westerly. It is a good example of a diverse, seasonally flooded palustrine deciduous scrub-shrub wetland.

Plate 15. Atlantic white cedar swamp (palustrine evergreen forested wetland) bordering Ell Pond, Rockville. A leatherleaf bog forms a floating mat along the shoreline.

Plate 16. Red maple swamp (palustrine deciduous forested wetland), Richmond. Most of these swamps occur in depressions, but some occur on seepage slopes.

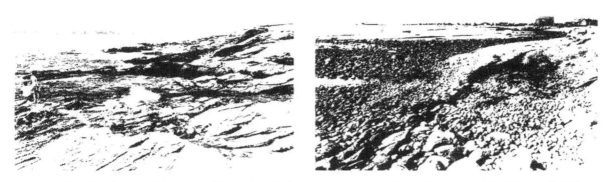

Figure 15. Marine intertidal wetlands form much of the Rhode Island coastline: (1) bedrock rocky shore at Beaver Tail (left) and (2) cobble-gravel unconsolidated shore at Point Judith (right).

within the salt marshes may be vegetated with widgeon grass (*Ruppia maritima*), sea lettuce (*Ulva lactuca*), or other algae.

The short form of smooth cordgrass forms extensive stands just above the low marsh. Within these and higher areas, shallow depressions called pannes can be found. These pannes are subjected to extreme temperatures and salinity. Summer salinities may exceed 40 parts per thousand (Martin 1959), while after heavy rains they may be filled with fresh water. Although they may be devoid of plants, many pannes are colonized by a short form of smooth cordgrass, glassworts, and spike grass, while blue-green algae may form a dense surface mat.

Figure 16. Estuarine emergent wetland and exposed tidal flat (low tide) at Succotash Marsh in Jerusalem.

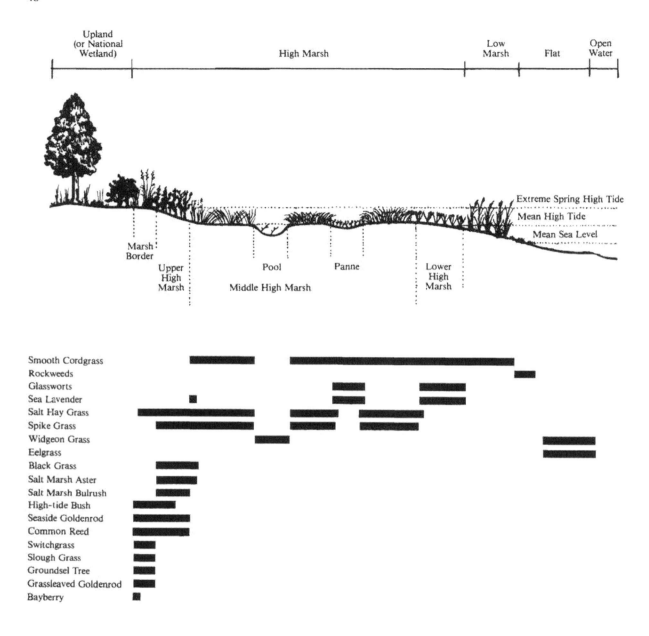

Figure 17. Generalized plant zonation in southern New England salt marshes: (1) low marsh and (2) high marsh. The high marsh can be further divided into several subzones. Note that plant diversity increases toward upland. (Source: adapted from Tiner 1987)

Above the short cordgrass marsh, two grasses and one rush predominate: salt hay grass, spike grass, and black grass (a member of the Rush Family). Salt hay grass often forms nearly pure stands, but it is frequently intermixed with spike grass. Spike grass usually forms pure or nearly pure stands in the more poorly drained high marsh areas where standing water is present for extended periods. The short form of smooth cordgrass also frequently occurs in this middle high marsh zone and is often intermixed with salt hay grass. Black grass, which is actually a rush, is found in abundance at slightly higher levels, often with high-tide bush (*Iva frutescens*). Other common high marsh plants include: seaside arrow grass (*Triglochin maritima*), salt marsh bulrush (*Scirpus robustus*), seaside plantain (*Plantago maritima*), sea lavender, marsh orach, salt marsh aster, seaside gerardia (*Agalinis maritima*), sea blite (*Suaeda maritima*), seaside goldenrod (*Solidago sempervirens*), and sand spurrey (*Spergularia maritima*).

Creeks and ditches throughout the high marsh are often immediately bordered by a tall or intermediate form of smooth cordgrass, while old spoil mounds adjacent to these mosquito ditches may be colonized by high-tide bush.

At the upland edge of salt marshes, switchgrass (*Panicum virgatum*), slough grass (*Spartina pectinata*), common reed (*Phragmites australis*), groundsel tree (*Baccharis halimifolia*), high-tide bush, and red cedar (*Juniperus virginiana*) frequently occur. Other plants present in border areas include bayberry (*Myrica pensylvanica*), poison ivy (*Toxicodendron radicans*), seaside goldenrod, grass-leaved goldenrod (*Euthamia graminifolia*), sweet grass (*Hierochloe odorata*), red fescue (*Festuca rubra*), wild rye (*Elymus virginicus*), purple loosestrife (*Lythrum salicaria*), and marsh pink (*Sabatia stellaris*). Where freshwater influence from the upland is strong, narrow-leaved cattail (*Typha angustifolia*), three-squares (*Scirpus americanus* and *S. pungens*), marsh fern (*Thelypteris thelypteroides*), rose mallow (*Hibiscus moscheutos*), creeping bent grass (*Agrostis stolonifera* var. *compacta*), spikerushes (*Eleocharis* spp.), Canada rush (*Juncus canadensis*), and other species may occur. These areas resemble brackish marshes which are more extensive in adjacent states.

Two Fish and Wildlife Service reports on New England salt marshes (Nixon 1982; Teal 1986) serve as useful references on the ecology of New England salt marshes.

Estuarine Scrub-Shrub Wetlands

Estuarine shrub wetlands are not common along the Rhode Island coast. Where present, they are usually dominated by high-tide bush. This shrub is especially common along mosquito ditches in salt marshes where it has become established on mounds of deposited material and along the upper edges of salt marshes. Salt hay grass, spike grass, and black grass are often co-dominants with high-tide bush in the high marsh. Groundsel tree may also be present along the upland border of salt marshes.

Estuarine Aquatic Beds

Shallow coastal embayments called "salt ponds" are a dominant feature along Rhode Island's coast. In Washington County, these ponds lie between the mainland and narrow barrier beaches. Most are connected to the ocean by manmade channels (breachways), while two ponds—Trustom Pond and Cards Ponds are not permanently connected to the sea. Ponds with breachways have higher salinities and are subjected to daily tidal action, while Trustom and Cards Ponds are brackish due to intermittent tidal influence. Prior to the building of permanent breachways, widgeon grass (*Ruppia maritima*) was the dominant aquatic bed plant. With increased salinities after breachway construction, eelgrass (*Zostera marina*) replaced much of the widgeon grass (Lee 1980). Other aquatic species reported in salt ponds include sago pondweed (*Potamogeton pectinatus*), clasping-leaved pondweed (*P. perfoliatus*), narrowleaf pondweed (*P. pusillus = P. berchtoldii*), horned pondweed (*Zannichellia palustris*), naiads (*Najas* spp.), and muskgrasses (*Chara* spp.) (Wright, *et al.* 1949). Thorne-Miller and others (1983) recently mapped the distribution of submerged aquatic bed vegetation in five of Rhode Island's coastal ponds.

Riverine Wetlands

The Riverine System encompasses all of Rhode Island's freshwater rivers and their tributaries, including the freshwater tidal reaches of coastal rivers where salinity is less than 0.5 ppt. This system is composed largely of deepwater habitats and nonvegetated wetlands, with the wetlands occurring between the riverbank and deep water (6.6 feet and greater in depth).

Although many of the state's freshwater vegetated wetlands lie along nontidal rivers and streams, only a small fraction of these are considered riverine wetlands according to the Service's classification system (Cowardin, *et al.* 1979). Riverine wetlands are by definition largely restricted to shallow bottoms and aquatic beds within the channels and to fringes of nonpersistent emergent plants growing on river banks or in shallow water. Contiguous wetlands dominated by persistent vegetation (i.e., trees, shrubs, and robust emergents) are classified as palustrine wetlands.

Riverine wetlands are most visible along slow-flowing, meandering lower perennial rivers and streams. Here nonpersistent emergent plants such as burreeds (*Sparganium* spp.), pickerelweed (*Pontederia cordata*), arrowheads (*Sagittaria* spp.), arrow arum (*Peltandra virginica*), wild rice (*Zizania aquatica*), rice cutgrass (*Leersia oryzoides*), true forget-me-not (*Myosotis scorpioides*), and smartweeds colonize very shallow waters and exposed shores. Aquatic beds may also become established in slightly deeper waters of clear rivers and streams. Important aquatic bed plants include submerged forms of burreeds and arrowheads, pondweeds and riverweeds (*Potamogeton* spp.), spatterdock (*Nuphar luteum*), and white water lily (*Nymphaea odorata*).

Palustrine Wetlands

Palustrine wetlands are the most common wetlands in Rhode Island. They represent the most floristically diverse group of wetlands in the state and include freshwater marshes, wet meadows, swamps, bogs, and shallow ponds. This collection of wetlands encompasses a wider range of water regimes than wetlands of other systems, with the more common water regimes being permanently flooded, semipermanently flooded, seasonally flooded, and saturated. Certain tidally influenced freshwater areas are also considered palustrine wetlands. While numerous plants may be restricted to one or two major hydrologic regimes, many plants, such as red maple (*Acer rubrum*) and purple loosestrife, tolerate a wide range of flooding and soil saturation conditions. Although their tolerances may be high, some wetland species are usually more prevalent under certain conditions and may, therefore, be used as indicators of certain water regimes. Examples of plant-water regime relationships are presented in Table 10 (Chapter 4). Palustrine wetland plant communities are discussed by class in the following subsections. The reader must recognize that due to the diversity of these communities, this discussion characterizes only the major types in general terms. Moreover, it is based largely on observations made during the inventory, since few studies have been conducted in Rhode Island's freshwater wetlands.

Palustrine Aquatic Beds

Small ponds, many of which were artificially-created, are common throughout the state. These permanently or semipermanently flooded water bodies comprise the wettest of palustrine wetlands. Many shallow ponds have aquatic beds covering all or part of their surfaces or bottoms. Common dominance types include green algae, floating species such as duckweeds (*Lemna* spp., *Spirodela polyrhiza*, and others) and bladderworts (*Utricularia* spp.), and rooted vascular plants, such as spatterdock, white water lily, water shield (*Brasenia schreberi*), watercress (*Nasturtium officinale*), and pondweeds.

Palustrine Emergent Wetlands

Palustrine emergent wetlands are primarily freshwater marshes and wet meadows dominated by persistent and nonpersistent grasses, rushes, sedges, and other herbaceous or grass-like plants. Nearly all of these wetlands are nontidal, with tidal freshwater marshes being virtually nonexistent in Rhode Island (F. Golet, pers. comm.). Plates 9–12 show examples of palustrine emergent wetlands.

Nontidal freshwater emergent wetlands are common throughout the state (Figure 18). Compared to the palustrine forested and shrub wetlands, however, these emergent wetlands represent only a small portion (three percent) of the state's freshwater wetlands. The water regime in freshwater marshes greatly affects plant community composition.

Semipermanently flooded emergent marshes may be dominated by broad-leaved cattail (*Typha latifolia*), bayonet rush (*Juncus militaris*), spatterdock, pickerelweed, common three-square (*Scirpus pungens*), arrow arum, water willow (*Decodon verticillatus*), and burreeds. Associated plants include duckweeds, white water lily, pondweeds, water parsnip (*Sium suave*), umbrella pennywort (*Hydrocotyle umbellata*), soft-stemmed bulrush (*Scirpus validus*), arrowheads, and a variety of submergent aquatics.

Dominant plants of seasonally flooded emergent wetlands include rice cutgrass, broad-leaved cattail, narrow-leaved cattail, common reed, soft rush (*Juncus effusus*), Canada rush (*J. canadensis*), water willow, reed canary grass (*Phalaris arundinacea*), bluejoint (*Calamagrostis canadensis*), sweet flag (*Acorus calamus*), arrow arum, manna grasses (*Glyceria canadensis* and others), wool grass (*Scirpus cyperinus*), three-way sedge (*Dulichium arundinaceum*), tussock sedge (*Carex stricta*), other sedges (e.g., *C. lurida*, *C. stipata*, and *C. vulpinoidea*), and grass-leaved goldenrod (*Euthamia graminifolia*). Soft rush and sweet flag are especially common in grazed wet meadows which may be seasonally flooded or saturated (Golet and Davis 1982). Sweet flag also occurs with broad-leaved cattail in seasonally flooded marshes. Other common plants are jewelweed (*Impatiens capensis*), arrow-leaved tearthumb (*Polygonum sagittatum*), purple loosestrife, smartweeds, sensitive fern (*Onoclea sensibilis*), marsh fern, boneset (*Eupatorium perfoliatum*), grasses (*Agrostis* sp. and others), green bulrush (*Scirpus atrovirens*), asters (*Aster* spp.), broad-leaved arrowhead (*Sagittaria latifolia*), spikerushes, Joe-Pye-weeds (*Eupatoriadelphus* spp.), bedstraw (*Galium tinctorium*), common three-square, swamp milkweed (*Asclepias incarnata*), and water purslane (*Ludwigia palustris*). Other emergents in seasonally flooded marshes and wet meadows include blue flag (*Iris versicolor*), bugleweeds (*Lycopus* spp.), marsh St. John's-wort (*Triadenum virginicum*), other St. John's-worts (*Hypericum* spp.), umbrella sedges (*Cyperus* spp.), false nettle (*Boehmeria cylindrica*), swamp dock (*Rumex verticillatus*), bittersweet nightshade (*Solanum dulcamara*), soft-stemmed bulrush, blue vervain (*Verbena hastata*), water parsnip, horsetail (*Equisetum* sp.), dodder (*Cuscuta gronovii*), swamp candles (*Lysimachia terrestris*), rough-stemmed goldenrod (*Solidago rugosa*), mad-dog skullcap (*Scutellaria lateriflora*), skunk cabbage (*Symplocarpus foetidus*), and panic grasses (*Panicum* spp. and *Dichantehlium* spp.). Shrubs, such as willows (*Salix* spp.), buttonbush (*Cephalanthus occidentalis*), multiflora rose (*Rosa multiflora*), swamp rose (*Rosa palustris*), northern arrowwood (*Viburnum recognitum*), common elderberry (*Sambucus*

Figure 18. Freshwater marsh near Woodville, dominated by tussock sedge.

canadensis), speckled alder (*Alnus rugosa*), steeplebush (*Spiraea tomentosa*), broad-leaved meadowsweet (*Spiraea latifolia*), highbush blueberry (*Vaccinium corymbosum*), and silky dogwood (*Cornus amomum*), and saplings of red maple may be scattered within these emergent wetlands.

Seasonally flooded emergent wetlands called "fens" are represented by an herbaceous community of woolly-fruit sedge (*Carex lasiocarpa*), twig-rush (*Cladium mariscoides*), bluejoint, and buckbean (*Menyanthes trifoliata*), with shrubs, mainly sweet gale (*Myrica gale*) and leatherleaf (*Chamaedaphne calyculata*), also common (Golet and Davis 1982). Emergent "bogs" carpeted with peat mosses (*Sphagnum* spp.) and having a saturated water regime are commonly dominated by beaked sedge (*Carex rostrata*), white beak-rush (*Rhynchospora alba*) or twig-rush, with cottongrass (*Eriophorum virginicum*) usually present (F. Golet, pers. comm.). Soft rush and Virginia chain fern (*Woodwardia virginica*) may also occur.

Temporarily flooded emergent wetlands may support soft rush, reed canary grass, common reed, goldenrods (*Solidago* spp. and *Euthamia* spp.), Joe-Pye-weeds, asters, and others. These emergent wetlands are less common than the seasonally flooded types.

Palustrine Scrub-shrub Wetlands

Scrub-shrub wetlands represent just less than 10 percent of Rhode Island's inland wetlands. They are characterized by the dominance of shrubs or tree saplings less than 20 feet tall. Two general types of scrub-shrub wetlands are present in the state: (1) deciduous scrub-shrub wetlands and (2) evergreen scrub-shrub wetlands, with the former group being most abundant. Mixtures of these types occur. Also, many shrub wetlands are intermixed with emergent wetlands. Examples of scrub-shrub wetland communities are presented in Table 17 and are pictured in Plates 13–14.

Deciduous shrub swamps may be dominated by one or more of several species including buttonbush, sweet gale, highbush blueberry, swamp azalea (*Rhododendron viscosum*), winterberries (*Ilex verticillata* and *I. laevigata*), northern arrowwood, alders (*Alnus rugosa* and *A. serrulata*), sweet pepperbush (*Clethra alnifolia*), silky dogwood, swamp rose, and saplings of red maple (*Acer rubrum*). The first two species usually occur in the wettest

Table 17. Examples of palustrine scrub-shrub wetlands in Rhode Island.

Dominance Type	Associated Vegetation	Location (County)
Highbush Blueberry	Common Winterberry, Swamp Azalea, Red Maple, Spikerush, Peat Moss, Tussock Sedge, Chokeberry, Serviceberry, Black Gum, White Pine, Cinnamon Fern, Canada Mayflower, Blue Flag, Gray Birch, Sheep Laurel, and Broad-leaved Meadowsweet	Providence
Buttonbush	Tussock Sedge, Alder, Bluejoint, Swamp Rose, and Broad-leaved Cattail	Kent
Well-Mixed Deciduous Shrubs	Red Maple, Ash, Northern Arrowwood, Willow, Smooth Winterberry, Swamp Rose, Multiflora Rose, Grape, Serviceberry, and Poison Ivy	Newport
Well-Mixed Deciduous Shrubs	Sweet Pepperbush, Highbush Blueberry, Swamp Azalea, Maleberry, Peat Mosses, Soft Rush, Twig-rush, Virginia Chain Fern, Red Maple, and Cottongrass	Newport
Highbush Blueberry/ Sweet Pepperbush	Mosses, Red Maple, Cinnamon Fern, White Pine, Ash, and Black Gum	Providence
Leatherleaf/ Tussock Sedge	Broad-leaved Meadowsweet, Peat Moss, Yellow Pond Lily, Steeplebush, Sweet Pepperbush, and Red Maple	Kent
Leatherleaf/Sweet Gale/ Sweet Pepperbush/Broad-leaved Meadowsweet	Tussock Sedge, Atlantic White Cedar, and White Pine	Kent
Red Maple (saplings)	Tussock Sedge, Bugleweed, Peat Moss, Blunt Manna Grass, and Sedge	Providence
Silky Dogwood	Swamp Rose, Speckled Alder, Red Maple, Joe-Pye-weed, Boneset, Sedge, Jewelweed, Sensitive Fern, Horsetail, Soft Rush, Northern Arrowwood, Grasses, Tussock Sedge, and Green Ash	Bristol
Red Maple	Highbush Blueberry, Common Winterberry, Sweet Pepperbush, Northern Arrowwood, Poison Ivy, Gray Birch, Cinnamon Fern, Sensitive Fern, and Skunk Cabbage	Bristol
Swamp Rose	Northern Arrowwood, Elderberry, Sensitive Fern, Skunk Cabbage, Broad-leaved Cattail, Red Maple, Serviceberry, Willow, and Reed Canary Grass	Newport
Well-Mixed Deciduous Shrubs	Sweet Gale, Swamp Rose, Sweet Pepperbush, Common Winterberry, Swamp Azalea, Highbush Blueberry, Poison Ivy, Black Chokeberry, Red Maple, Peat Moss, and Marsh Fern	Washington
Highbush Blueberry/ Water Willow/ Swamp Azalea	Cinnamon Fern, Sweet Pepperbush, Peat Moss, White Pine, Skunk Cabbage, Serviceberry, Canada Mayflower, Wild Calla, and Maleberry; Edges: Red Maple, Blue Flag, Broad-leaved Meadowsweet, and Gray Birch	Providence
Speckled Alder/Willow/ Common Winterberry	Red Maple, Buttonbush, Swamp Rose, and Steeplebush with various emergents	Providence

areas, often in shallow water along the shores of lakes, ponds or rivers, whereas the other species are generally associated with seasonally flooded wetlands (Figure 19). Water willow may be intermixed with buttonbush in many shallow water areas. Leatherleaf is sometimes an associate of sweet gale (Golet and Davis 1982). Wright (1941) observed royal fern (*Osmunda regalis*), northern wild raisin (*Viburnum cassinoides*), leatherleaf, maleberry (*Lyonia ligustrina*), and hoary willow (*Salix candida*) in mixed deciduous shrub swamps within the Great Swamp in Washington County. Other common shrubs include oblong-leaf shadbush (*Amelanchier canadensis*), steeplebush, broad-leaved meadowsweet, chokeberries (*Aronia* spp.), common elderberry (*Sambucus canadensis*), poison ivy (*Toxicodendron radicans*), and willows (*Salix* spp.). Occasionally bayberry (*Myrica pensylvanica*) and gray birch (*Betula populifolia*) are found in palustrine wetlands. A variety of herbaceous plants are common in shrub swamps. Peat mosses (*Sphagnum* spp.) may be present in great abundance, especially in boggy areas.

Evergreen scrub-shrub wetlands are represented by two types of communities: (1) leatherleaf bogs and (2) Atlantic white cedar sapling swamps. The former type is dominated by leatherleaf, with associated species including pitcher plant (*Sarracenia purpurea*), rose pogonia (*Pogonia ophioglossoides*), round-leaved sundew (*Drosera rotundifolia*), cotton-grasses (*Eriophorum virginicum* and *E. tennellum*), and bog clubmoss (*Lycopodium inundatum*) (Wright 1941). Sweet gale, sweet pepperbush, broad-leaved meadowsweet, tussock sedge, white pine (*Pinus strobus*), and Atlantic white cedar (*Chamaecyparis thyoides*) may also be present. Golet and Davis (1982) observed peat mosses, black huckleberry (*Gaylussacia baccata*), sheep laurel (*Kalmia angustifolia*), and cranberries (*Vaccinium macrocarpon* and *V. oxycoccus*) in this wetland type. Other bog species include white

Figure 19. Palustrine scrub-shrub wetland in Chapman Swamp in Westerly.

beak-rush, dwarf huckleberry (*Gaylussacia dumosa*), water willow, grass pink (*Calopogon tuberosus*), and Dragon's Mouth (*Arethusa bulbosa*) (F. Golet, pers. comm.). Wright (1941) described a regenerating Atlantic white cedar swamp in the Great Swamp. Plants associated with Atlantic white cedar in this swamp included rose bay (*Rhododendron maximum*), mountain holly (*Nemopanthus mucronata*), poison sumac (*Toxicodendron vernix*), grass pink, rose pogonia, round-leaved sundew, bunchberry (*Cornus canadensis*), clintonia (*Clintonia borealis*), sweet white violet (*Viola blanda*), goldthread (*Coptis trifolia*), white fringed orchis (*Platanthera blephariglottis*), and pitcher plant. Leatherleaf and sweet gale may also be present in this type of wetland (F. Golet, pers. comm.).

Palustrine Forested Wetlands

Palustrine forested wetlands are the state's most abundant wetland type, representing about 73 percent of the state's wetlands and 83 percent of the nontidal wetlands. These wetlands are common along rivers and streams, in isolated depressions, and in drainageways on hillsides. Forested wetlands are characterized by the dominance of woody plants 20 feet or taller. Three general types of forested wetlands occur in Rhode Island: (1) broad-leaved deciduous forested wetlands, (2) needle-leaved evergreen forested wetlands, and (3) mixtures of deciduous and evergreen forested wetlands. The former type is by far the more abundant type. Most of the state's forested wetlands are seasonally flooded; hillside swamps are saturated, while temporarily flooded wetlands are uncommon. Table 18 presents examples of forested wetland plant communities in Rhode Island, while Plates 15–16 illustrate two of them.

The majority of the broad-leaved deciduous forested wetlands are red maple swamps. Red maple dominates these wetlands, with numerous other plants occurring as associates. Common subordinates in these swamps include black gum (*Nyssa sylvatica*), white pine, Atlantic white cedar, and hemlock (*Tsuga canadensis*). White pine is frequently co-dominant. Other trees may also be present in lesser abundance, such as yellow birch (*Betula alleghaniensis*), gray birch, green ash (*Fraxinus pennsylvanica*), white ash (*Fraxinus americana*), American elm (*Ulmus americana*), pitch pine (*Pinus rigida*), black cherry (*Prunus serotina*), oblong-leaf shadbush, swamp

Table 18. Examples of palustrine forested wetlands in Rhode Island.

Dominance Type	Associated Vegetation		Location (County)
Red Maple	Trees:	American Elm, Ironwood	Providence
	Shrubs:	Spicebush, Northern Arrowwood, Japanese Barberry, Northern Wild Raisin	
	Herbs:	Skunk Cabbage, Wood Anemone, False Hellebore, Canada Mayflower, Tall Meadow-rue, Jewelweed, Sedge, Aster	
Red Maple	Shrubs:	Sweet Pepperbush, Northern Arrowwood, Elderberry, Highbush Blueberry, Northern Wild Raisin	Providence
	Herbs:	Marsh Marigold, Skunk Cabbage, Tussock Sedge	
	Others:	Common Greenbriar	
Red Maple	Trees:	Yellow Birch, Black Gum, White Pine	Providence
	Shrubs:	Highbush Blueberry, Sweet Pepperbush	
	Herbs:	Skunk Cabbage, Cinnamon Fern, Canada Mayflower, Blue Flag	
	Others:	Peat Moss, other Mosses	
Red Maple	Shrubs:	Sweet Pepperbush, Spicebush, Elderberry, Serviceberry	Providence
	Herbs:	Jewelweed, Skunk Cabbage, Cinnamon Fern, Grass, Canada Mayflower, Wood Anemone, Turk's-cap Lily, Northern White Violet, Tall Meadow-rue, Horsetail, False Hellebore, Goldenrod	
Red Maple	Trees:	Atlantic White Cedar, White Oak	Washington
	Shrubs:	Rose Bay, Highbush Blueberry, Common Winterberry	
	Herbs:	Cinnamon Fern	
Red Maple	Shrubs:	Rose Bay, Swamp Azalea, Sweet Pepperbush	Washington
	Herbs:	Skunk Cabbage, Jack-in-the-pulpit, Cinnamon Fern	
	Others:	Peat Mosses, Swamp Dewberry	
Red Maple	Shrubs:	Sweet Pepperbush, Common Winterberry, Swamp Azalea	Newport
	Herbs:	Skunk Cabbage, Sensitive Fern, Royal Fern, Sedge	
	Others:	Greenbriar, Poison Ivy	
Red Maple/ Swamp White Oak/ American Elm	Shrubs:	Northern Arrowwood, Highbush Blueberry, Common Winterberry, Swamp Rose	Bristol
	Herbs:	Sensitive Fern, Grass, Skunk Cabbage, Jewelweed, Clearweed, Blue Flag, Crowfoot	
	Others:	Poison Ivy	
Red Maple/White Pine	Shrubs:	Highbush Blueberry, Sweet Pepperbush, Northern Wild Raisin, Swamp Azalea, Broad-leaved Meadowsweet, Chokeberry, Elderberry, Sheep Laurel, Mountain Holly	Kent
	Herbs:	Tussock Sedge, Skunk Cabbage, Cinnamon Fern, Goldthread, Turk's-cap Lily, Tall Meadow-rue, Sensitive Fern, Canada Mayflower, Sedge, Violet, Goldenrod	
	Others:	Peat Mosses, Grape	
Red Maple/White Pine	Trees:	Yellow Birch, Ash	Providence
	Shrubs:	Alder, Northern Arrowwood, Sweet Pepperbush, Swamp Azalea, Silky Dogwood, Highbush Blueberry, Black Cherry, Northern Wild Raisin, Maleberry, Broad-leaved Meadowsweet, Swamp Rose, Smooth Gooseberry	
	Herbs:	Skunk Cabbage, Wood Anemone, Canada Mayflower, Cinnamon Fern, Tall Meadow-rue, False Hellebore, Sedge, Bellflower, Goldenrod, Bluejoint, Sensitive Fern, Whorled Aster, Marsh Blue Violet	
Red Maple/White Pine	Trees:	Yellow Birch, White Oak	Newport
	Shrubs:	Alder, Highbush Blueberry, Common Winterberry, Sweet Pepperbush, Sheep Laurel, Northern Arrowwood	
	Herbs:	Cinnamon Fern, Aster	
	Others:	Peat Mosses, Grape	
Red Maple/White Pine	Trees:	Pitch Pine	Providence
	Shrubs:	Sheep Laurel, Highbush Blueberry, Swamp Azalea, Northern Arrowwood, Broad-leaved Meadowsweet, Common Winterberry	
	Herbs:	Ground Pine, Canada Mayflower, Cinnamon Fern, Marsh St. John's-wort, Sedge, Royal Fern, Skunk Cabbage, Three-way Sedge	
	Others:	Peat Mosses, Greenbriar	

Table 18. (Continued)

Dominance Type	Associated Vegetation		Location (County)
Red Maple/ Atlantic White Cedar	Trees:	Yellow Birch, White Pine	Washington
	Shrubs:	Highbush Blueberry, Swamp Azalea, Rose Bay, Northern Wild Raisin, Sweet Pepperbush	
	Herbs:	Cinnamon Fern, Skunk Cabbage, Starflower, Marsh Fern, Wild Sarsaparilla	
	Others:	Peat Mosses	
Red Maple/Hemlock	Trees:	Gray Birch, White Pine	Providence
	Shrubs:	Highbush Blueberry, Northern Wild Raisin, Winterberry, Broad-leaved Meadowsweet, Spicebush, Chokeberry, Maleberry, Sheep Laurel	
	Herbs:	Skunk Cabbage, Turk's-cap Lily, Blue Flag, Northern White Violet, Goldenrod or Aster, Tussock Sedge	
	Others:	Peat Mosses, other Mosses, Swamp Dewberry	
White Pine/Red Maple	Trees:	Atlantic White Cedar, Ash, Red Oak	Washington
	Shrubs:	Sweet Pepperbush, Spicebush, Common Winterberry, Swamp Azalea, Poison Sumac	
	Herbs:	Skunk Cabbage, New York Fern, Royal Fern, Shining Clubmoss	
	Others:	Ground Strawberry	
White Pine/Red Maple	Shrubs:	Highbush Blueberry, Sweet Pepperbush, Swamp Azalea, Inkberry, Common Winterberry, Northern Wild Raisin	Washington
	Herbs:	Cinnamon Fern, Skunk Cabbage, Ground Pine, Starflower	
	Others:	Peat Mosses, Partridgeberry	
White Pine	Trees:	Red Maple, White Oak	Kent
	Shrubs:	Sweet Pepperbush, Northern Wild Raisin, Highbush Blueberry, Sheep Laurel, Swamp Azalea, Wintergreen	
	Herbs:	Cinnamon Fern, Skunk Cabbage	
	Others:	Common Greenbriar, Peat Mosses, other Mosses, Ground Pine, Partridgeberry	
Atlantic White Cedar	Trees:	Hemlock, Red Maple, White Pine, Red Oak	Kent
	Shrubs:	Highbush Blueberry	
	Herbs:	Cinnamon Fern, Skunk Cabbage	
	Others:	Peat Mosses	
Atlantic White Cedar	Trees:	Red Maple, Black Gum	Washington
	Shrubs:	Sweet Pepperbush, Highbush Blueberry, Swamp Azalea, Mountain Laurel	
	Others:	Peat Mosses, other Mosses	
Hemlock	Trees:	Yellow Birch, Black Gum, Red Maple	Providence
	Shrubs:	Mountain Laurel, Spicebush, Alder	
	Herbs:	Skunk Cabbage, Sedge, Marsh Marigold, False Hellebore	
	Others:	Peat Mosses, other Mosses	

white oak (*Quercus bicolor*), and ironwood (*Carpinus carolinianus*). Red maple swamps typically have a dense shrub understory comprised mostly of sweet pepperbush, highbush blueberry, swamp azalea, rose bay, northern arrowwood, common winterberry (*Ilex verticillata*), smooth winterberry (*I. laevigata*), swamp sweetbells (*Leucothoe racemosa*), and spicebush (*Lindera benzoin*) (Figure 20). Less common shrubs that may be locally common include maleberry, northern wild raisin, meadowsweet, silky dogwood, black chokecherry (*Aronia melanocarpa*), swamp rose, alder buckthorn (*Rhamnus frangula*), sheep laurel, and common elderberry. Witch hazel (*Hamamelis virginiana*) may occur along the upper edges of these swamps. Several vines have been observed in Rhode Island red maple swamps, including poison ivy, common greenbriar (*Smilax rotundifolia*), sawbriar (*S. glauca*), climbing hempweed (*Mikania scandens*), ground-nut (*Apios americana*), grape (*Vitis* sp.), and Virginia creeper (*Parthenocissus quinquefolia*) (personal observations; F. Golet, pers. comm.). The herbaceous layer in these swamps is variable, with areal coverage ranging from 0.2 percent to 70 percent in six red maple swamps studied by Lowry (1984). Common plants in this layer are cinnamon fern (*Osmunda cinnamomea*), sensitive fern, royal fern, marsh fern, Massachusetts fern (*Thelypteris simulata*), jewelweed, tussock and other sedges, skunk cabbage, violets (*Viola pallens* and *V. cucullata*), netted chain fern (*Woodwardia areolata*), marsh marigold (*Caltha palustris*), and water horehounds (*Lycopus* spp.). Less common plants include crested fern (*Dryopteris cristata*), wood reed (*Cinna arundinacea*), blue flag, cardinal flower (*Lobelia cardi-*

Figure 20. Red maple swamps are the predominant wetland type in the state. Rose bay is a common understory species.

nalis), tall meadow-rue (*Thalictrum pubescens*), Turk's-cap lily (*Lilium canadense* ssp. *michiganense*), false hellebore (*Veratrum viride*), jack-in-the-pulpit (*Arisaema triphyllum*), clearweed (*Pilea pumila*), and turtlehead (*Chelone glabra*). At higher levels in the swamp, Canada mayflower (*Maianthemum canadense*), ground pine (*Lycopodium obscurum*), shining clubmoss (*Lycopodium lucidulum*), partridgeberry (*Mitchella repens*), wild sarsaparilla (*Aralia nudicaulis*), painted trillium (*Trillium undulatum*), starflower (*Trientalis borealis*), and wood anenome (*Anenome quinquefolia*) may be present (personal observations; Wright 1941). Peat mosses form abundant groundcover in depressions within many seasonally flooded and saturated red maple swamps. Swamp dewberry (*Rubus hispidus*) is also a common groundcover plant. A study on plant-soil relationships along the transition zone of Rhode Island's red maple swamps has been completed (Allen, *et al.* 1989). Davis (1988) examined vegetation-hydrology gradients in red maple swamps. In the near future, a Fish and Wildlife Service report on the ecology of red maple swamps will be published (Golet, *et al.* 1990).

Wright (1941) reported a mixed deciduous forested wetland community in the Great Swamp. This community was slightly higher in elevation than the adjacent red maple swamp community. This mixed type was represented by several species of oak, including white oak (*Quercus alba*), black oak (*Q. velutina*), red oak (*Q. rubra*), swamp white oak (*Q. bicolor*), and pin oak (*Q. palustris*), and other trees, such as red maple, yellow birch, white ash, black gum, and white pine. American holly (*Ilex opaca*) was quite common in this community, while it was only scattered throughout other forested wetland types in the Great Swamp.

Evergreen forested wetlands are dominated by one of three conifers: (1) Atlantic white cedar, (2) white pine, and (3) hemlock. While most people associate white cedar with wetlands, many people think of white pine as an upland plant. Although white pine appears to be prevalent on well-drained upland sites, it is also quite common in Rhode Island's wetlands. In fact, Bromley (1935), in his discussion of the original forest types of southern New England, reported that at the time of settlement white pine was probably abundant only in swamps and moist sandy pine flats, and on exposed ridges. This distribution was attributed to white pine's susceptibility to fire. Today, wildfires are greatly suppressed, so white pine can thrive on better drained sites than it could originally. Surprisingly, however, Mader (1976) in studying growth and productivity of white pine in Massachusetts found that the better sites were associated with the poorer drainage classes of soils (e.g., poorly drained and somewhat poorly drained).

Atlantic white cedar swamps are not as common as they were in the past. Many cedar swamps have been replaced by hardwood swamps or white pine-hemlock swamps (Bromley 1935). Most of the swamps called "Cedar Swamp" in Rhode Island (e.g., in Woonsocket, West Greenwich, and Wood River Junction) on the U.S.G.S. topographic maps are now red maple swamps. Atlantic white cedar swamps are most abundant in Washington County and the western parts of Kent and Providence Counties (Laderman, et al. 1987). They are associated with the state's largest wetlands, namely Chapman Swamp (Westerly), Indian Cedar Swamp (Charlestown), and Great Swamp (South Kingston, Richmond, and Charlestown). In the remaining cedar swamps, red maple appears to be the most common tree associated with the cedars, whereas black gum, gray birch, yellow birch, black birch (*Betula lenta*), white oak, white pine, hemlock, and pitch pine are less abundant. Hemlock may be common in the understory in some cedar swamps. In a few swamps in northwestern Rhode Island, Atlantic white cedar is associated with black spruce (*Picea mariana*) and larch (*Larix laricina*) (Laderman, et al. 1987). Lowry (1984) found an average of 16 shrub species growing in six Atlantic white cedar swamps in southern Rhode Island. Common shrubs included sweet pepperbush, rose bay, swamp azalea, highbush blueberry, swamp sweetbells, common winterberry, and smooth winterberry. Sweet pepperbush was the most abundant shrub. Mountain holly, inkberry (*Ilex glabra*), and sheep laurel may also be present in lesser amounts. The herbaceous stratum generally is not as extensive as in red maple swamps, except where the tree canopy is more open. Lowry (1984) observed herbaceous covers ranging to 29 percent, with most of the study swamps having less than 4 percent coverage. This low coverage resulted from dense shading by the cedars and the shrub understory. Common herbaceous plants in cedar swamps include water willow, broad-leaved arrowhead, sedges (*Carex trisperma, C. rostrata,* and *C. lasiocarpa*), cinnamon fern, marsh fern, and starflower. Other herbaceous plants of interest include round-leaved sundew, pitcher plant, bladderworts, buckbean, marsh St. John's-wort, arrow arum, grass pink, Massachusetts fern, and cotton-grasses. Canada mayflower and creeping wintergreen (*Gaultheria hispidula*) may be found on the hummocks at the bases of the cedars, while wild calla (*Calla palustris*) or marsh marigold (*Caltha palustris*) may occur in shaded pools (personal observations; Bromley 1935). Peat mosses and other mosses are common groundcover plants in cedar swamps.

White pine swamps are more abundant in central and northern Rhode Island which lie within the White Pine Region of southern New England characterized by Bromley (1935). As in the cedar swamps, red maple is the most common tree intermixing with white pine to varying degrees. In many areas, red maple often co-dominates, forming mixed evergreen/deciduous forested wetlands. Less common tree associates include Atlantic white cedar, ash, red oak, white oak, and yellow birch. The shrub layer contains several common species, including sweet pepperbush, highbush blueberry, swamp azalea, spicebush, and rose bay. Other shrubs observed in lesser amounts are winterberry, wintergreen (*Gaultheria procumbens*), northern wild raisin, sheep laurel, mountain laurel (*Kalmia latifolia*), inkberry, poison sumac, and partridgeberry. Herbaceous plants, such as skunk cabbage, goldthread (*Coptis trifolia*), cinnamon fern, royal fern, New York fern (*Thelypteris noveboracensis*), ground pine, shining clubmoss, and starflower may be present. Peat mosses may be found in varying amounts.

Hemlock swamps are less common than the other two evergreen forested wetland types and are relatively uncommon in Rhode Island. One such swamp observed in Foster had yellow birch, skunk cabbage, peat mosses, and other mosses as common associates. Plants in lesser amounts included mountain laurel, black gum, red maple, sedges, and marsh marigold. Contiguous hemlock areas were co-dominated by red maple and white pine.

Lacustrine Wetlands

The Lacustrine System is principally a deepwater habitat system of freshwater lakes, reservoirs and deep ponds. Consequently, wetlands are generally limited to shallow waters and exposed shorelines, as in the Riverine System. While algae are probably more abundant in these waters, vascular macrophytes are often more conspicuous. A variety of life forms can be recognized: (1) free-floating plants, (2) rooted vascular floating-leaved plants, (3) submergent plants, and (4) nonpersistent emergent plants. The first three life forms characterize aquatic beds, whereas the latter dominates lacustrine emergent wetlands.

Lacustrine Aquatic Beds

Floating-leaved and free-floating aquatic beds are common in shallow lacustrine waters. Common floating-leaved plants include white water lily, spatterdock, water shield, pondweeds, and floating heart (*Nymphoides aquatica*). Duckweeds (*Lemna* spp. and *Spirodela polyrhiza*)

Figure 21. Lacustrine nonpersistent emergent wetland along Worden Pond in South Kingston. Bayonet rush predominates.

comprise free-floating aquatic beds, while bladderworts are also free-floating, but are typically submerged. Submergent aquatic beds may include pondweeds, bushy pondweeds (*Najas* spp.), wild celery (*Vallisneria americana*), waterweeds (*Elodea* spp.), water milfoils (*Myriophyllum* spp.), mermaidweed, coontail (*Ceratophyllum demersum*), and fanwort (*Cabomba caroliniana*).

Lacustrine Emergent Wetlands

Emergent wetlands commonly border the margins of lakes, reservoirs, and deep ponds (Figure 21). Common nonpersistent emergent plants may include bayonet rush, common three-square, yellow-eyed grass (*Xyris* sp.), pipeworts (*Eriocaulon* spp.), pickerelweed, burreeds, arrowheads, water parsnip, three-way sedge, and spikerushes. Some of these plants are usually persistent, but along lake shores are subject to ice-scouring and therefore, may be considered nonpersistent. In many areas, persistent plants like cattails, reed canary grass, bluejoint, water willow, buttonbush, sweet gale, leatherleaf, swamp rose and others may form part of the lacustrine boundary. These persistent plants, however, represent palustrine wetlands along the lake shore.

References

Allen, S.D., F.C. Golet, A.F. Davis, and T.E. Sokoloski. 1989. Soil-vegetation Correlations in Transition Zones of Rhode Island Red Maple Swamps. U.S. Fish and Wildlife Service, Washington, DC. Biological Report 89(8). 47 pp.

Bromley, S.W. 1935. The original forest types of southern New England. Ecol. Monog. 5(1): 61–89.

Cowardin, L.M., V. Carter, F.C. Golet, and E.T. LaRoe. 1979. Classification of Wetlands and Deepwater Habitats of the United States. U.S. Fish and Wildlife Service. FWS/OBS-79/31. 103 pp.

Davis, A.F. 1988. Hydrologic and Vegetation Gradients in the Transition Zone of Rhode Island Red Maple Swamps, M.S. thesis, Natural Resources, University of Rhode Island, Kingstown, 136 pp. plus appendices.

Golet, F.C. and A.F. Davis. 1982. Inventory and Habitat Evaluation of the Wetlands of Richmond, Rhode Island. University of Rhode Island, College of Resource Development, Kingston. Occas. Paper in Env'tal Science No. 1. 48 pp.

Golet, F.D., A. Calhoun, D. Lowry, W. DeRagon, and A.J. Gold. 1990 (in press). Ecology of Red Maple Forested Wetlands in the Glaciated Northeastern United States. U.S. Fish and Wildlife Service, Washington, DC. Biological Report.

Kennard, W.C., M.W. Lefor, and D.L. Civco. 1983. Analysis of Coastal Marsh Ecosystems: Effects of Tides on Vegetational Change. Univ. of Connecticut, Institute of Water Resources, Storrs. Res. Proj. Tech. Completion Rept. B-014 CONN. 110 pp.

Laderman, A.D., F.C. Golet, B.A. Sorrie, and H.L. Woolsey. 1987.

Atlantic white cedar in the Glaciated Northeast. *In:* A.D. Laderman (editor). Atlantic White Cedar Wetlands. Westview Press, Boulder, CO. pp. 19–34.

Lee, V. 1980. An Elusive Compromise: Rhode Island Coastal Ponds and Their People. University of Rhode Island, Coastal Resources Center. Narragansett. Mar. Tech. Report. 73. 82 pp.

Lowry, D.J. 1984. Water Regimes and Vegetation of Rhode Island Forested Wetlands. M.S. Thesis, Plant and Soil Science. University of Rhode Island, Kingston. 174 pp.

Mader, D.L. 1976. Soil-site productivity for natural stands of white pine in Massachusetts. Soil Science Society of America Journal 40(1): 112–115.

Magee, D.W. 1981. Freshwater Wetlands: A Guide to Common Indicator Plants of the Northeast. University of Massachusetts Press, Amherst. 246 pp.

Martin, W.E. 1959. The vegetation of Island Beach State Park, New Jersey. Ecol. Monogr. 29(1): 1–46.

Nixon, S.W. 1982. The Ecology of New England High Salt Marshes: A Community Profile. U.S. Fish and Wildlife Service. FWS/OBS-81/55. 70 pp.

Odum, W.E., T.J. Smith III, J.K. Hoover, and C.C. McIvor. 1984. The Ecology of Tidal Freshwater Marshes of the United States East Coast: A Community Profile. U.S. Fish and Wildlife Service. FWS/OBS-83/17. 177 pp.

Penfound, W.T. 1952. Southern swamps and marshes. Bot. Rev. 18: 413–446.

Reed, P.B., Jr. 1988. National List of Plant Species that Occur in Wetlands: 1988 National Summary. U.S. Fish and Wildlife Service, National Ecology Center, Ft. Collins, CO. Biol. Rep. 88(24). 244 pp.

Simpson, R.L., R.E. Good, M.A. Leck, and D.F. Whigham. 1983. The ecology of freshwater tidal wetlands. BioScience 33: 255–259.

Thorne-Miler, B., M.M. Harlin, G.B. Thursby, M.M. Brady-Campbell, and B.A. Dworetzky. 1983. Variations in the distribution and biomass of submerged macrophytes in five coastal lagoons in Rhode Island, U.S.A. Bot. Mar. 26:12.

Tiner, R.W., Jr. 1987. A Field Guide to Coastal Wetland Plants of the Northeastern United States. University of Massachusetts Press, Amherst. 286 pp.

Tiner, R.W., Jr. 1988a. The occurrence of eastern white pine in New England wetlands. Massachusetts Assoc. of Conservation Commissions, Medford. MACC Newsletter XVII(6):10.

Tiner, R.W., Jr. 1988b. Field Guide to Nontidal Wetland Identification. Maryland Dept. of Natural Resources, Annapolis, MD and U.S. Fish and Wildlife Service, Newton Corner, MA. Cooperative publication. 283 pp. + 198 color plates.

U.S.D.A. Soil Conservation Service. 1982. National List of Scientific Plant Names. Vol I. List of Plant Names. SCS-TP-159. 416 pp.

U.S. Fish and Wildlife Service. 1982. Preliminary list of hydrophytes for the northeastern United States. Unpublished mimeo.

Whitlatch, R.B. 1982. The Ecology of New England Tidal Flats: Community Profile. U.S. Fish and Wildlife Service. FWS/OBS-81/01. 125 pp.

Wright, K.E. 1941. The Great Swamp. Torreya 41(5): 145–150.

Wright, T.J., V.I. Cheadle, and E.A. Palmatier. 1949. A Survey of Rhode Island's Salt and Brackish Water Ponds and Marshes. Rhode Island Department of Agriculture and Conservation, Division of Fish and Game, Providence. Pittman Robertson Pamphlet No. 2. 42 pp.

CHAPTER 7.
Wetland Values

Introduction

Rhode Island's wetlands have been traditionally used for hunting, trapping, fishing, berry harvest, timber and salt hay production, and livestock grazing. These uses tend to preserve the wetland integrity, although the qualitative nature of wetlands may be modified, especially by salt hay production and timber harvest. Human uses are not limited to these activities, but also include destructive and often irreversible actions such as drainage for agriculture and filling for industrial or residential development. In the past, many people considered wetlands as wastelands whose best use could only be attained through "reclamation projects" which led to the destruction of many wetlands. To the contrary, wetlands in their natural state provide a wealth of values to society (Table 19). These benefits can be divided into three basic categories: (1) fish and wildlife values, (2) environmental quality values, and (3) socio-economic values. The following discussion emphasizes the more important values of Rhode Island's wetlands, with significant national examples also presented. For an in-depth examination of wetland values, the reader is referred to *Wetland Functions and Values: The State of Our Understanding* (Greeson, *et al.* 1979). In addition, the U.S. Fish and Wildlife Service has created and maintains a wetland values database which records abstracts for over 5000 articles.

Table 19. List of major wetland values.

Fish and Wildlife Values

- Fish and Shellfish Habitat
- Waterfowl and Other Bird Habitat
- Mammal and Other Wildlife Habitat

Environmental Quality Values

- Water Quality Maintenance
  - Pollution Filter
  - Sediment Removal
  - Oxygen Production
  - Nutrient Recycling
  - Chemical and Nutrient Absorption
- Aquatic Productivity
- Microclimate Regulator
- World Climate (Ozone layer)

Socio-economic Values

- Flood Control
- Wave Damage Protection
- Shoreline Erosion Control
- Ground-water Recharge
- Water Supply
- Timber and Other Natural Products
- Energy Source (Peat)
- Livestock Grazing
- Fish and Shellfishing
- Hunting and Trapping
- Recreation
- Aesthetics
- Education and Scientific Research

Fish and Wildlife Values

Fish and wildlife utilize wetlands in a variety of ways. Some animals are entirely wetland-dependent, spending their entire lives in wetlands. Others use wetlands only for specific reasons, such as reproduction and nursery grounds, feeding, and resting areas during migration. Many upland animals visit wetlands to obtain drinking water and food. In urbanizing areas, the remaining wetlands become important habitats—a type of refuge—for "upland" wildlife displaced by development (F. Golet, pers. comm.). Wetlands are also essential habitat for numerous rare and endangered animals and plants.

Fish and Shellfish Habitat

Due to their linkage with adjacent waters, Rhode Island's coastal and inland wetlands are important fish habitats. Estuarine wetlands are also essential habitats for grass shrimp, crabs, oysters, clams, and other invertebrates.

Approximately two-thirds of the major U.S. commercial fishes depend on estuaries and salt marshes for nursery or spawning grounds (McHugh 1966). Among the more familiar wetland-dependent fishes are menhaden, bluefish, flounder, white perch, sea trout, mullet, croaker, striped bass, and drum. Forage fishes, such as anchovies, killifishes, mummichogs, and Atlantic silversides, are among the most abundant estuarine fishes. Narragansett Bay and its associated wetlands are important spawning and nursery grounds for many fish species (T. Lynch, pers. comm.). Winter flounder spawn in the shallow shoals of the Bay on beds of sea lettuce (*Ulva lactuca*), with peak spawning taking place from January to March. These same beds are used in the spring by spawning tautogs. Other nearshore spawners include scup, butterfish, and squid. Coastal ponds serve as spawning areas for tomcod beginning in November. As many as 63 fish species use Narragansett Bay as a nursery ground, with highest use in the fall.

Coastal wetlands are also important for shellfish including bay scallops, grass shrimp, blue crabs, oysters, quahogs and other clams. A critical stage of the bay scallop's life cycle requires that larvae attach to eelgrass leaves for about a month (Davenport 1903). Blue crabs and grass shrimp are abundant in tidal creeks of salt marshes. Estuarine aquatic beds, in general, also provide important cover for juvenile fishes and other estuarine organisms (Good, *et al.* 1978).

Figure 22. Migratory birds depend on Rhode Island wetlands: (a) black duck, (b) Canada goose goslings, (c) American bittern, and (d) yellow warbler.

Freshwater fishes also find wetlands essential for survival. In fact, nearly all freshwater fishes can be considered wetland-dependent because: (1) many species feed in wetlands or upon wetland-produced food, (2) many fishes use wetlands as nursery grounds, and (3) almost all important recreational fishes spawn in the aquatic portions of wetlands (Peters, *et al.* 1979). Many rivers and streams along Rhode Island's coast are spawning grounds for alewife and a few rivers are also used by sea-run brown trout, rainbow smelt, and American shad. Common fishes in Rhode Island's freshwater rivers, lakes, and ponds include northern pike, chain pickerel, largemouth bass, smallmouth bass, bluegill, common sunfish, yellow perch, brown bullhead, brook trout, rainbow trout, and white perch (Guthrie and Stolgitis 1977; RI DEM, pers. comm.). Northern pike spawn in early spring in flooded marshes and aquatic beds, while chain pickerel prefer aquatic beds. White perch are also early spring spawners, spawning in ponds and brackish coastal waters. Smallmouth bass spawn in about two feet of water from late May to early June. For all fish species, the presence of aquatic vegetation helps juvenile fishes avoid predator attacks, so wetlands are important nursery grounds.

Waterfowl and Other Bird Habitat

In addition to providing year-round habitats for resident birds, wetlands are particularly important as breeding grounds, over-wintering areas and feeding grounds for migratory waterfowl and numerous other birds (Figure 22). Both coastal and inland wetlands are valuable bird habitats.

Rhode Island's salt marshes are used for nesting by birds such as common terns, clapper rails, king rails, mallards, black ducks, blue-winged teals, mute swans, willets, herring gulls, great black-backed gulls, red-winged blackbirds, marsh wrens, sharp-tailed sparrows, and seaside sparrows. Red-winged blackbirds and seaside sparrows prefer stands of the short form of smooth cordgrass (*Spartina alterniflora*) which border permanent salt ponds, while marsh wrens prefer stands of the tall form of smooth cordgrass bordering tidal creeks and ditches (Reinert, *et al.* 1981). Moreover, the availability of open water and/or the short form smooth cordgrass community are directly related to the density of all breeding species. Bird breeding densities are over 2.5 times higher in unditched salt marshes than in ditched marshes (Reinert, et al. 1981). Wading birds, such as little blue herons, black-crowned night herons, glossy ibises, cattle egrets, snowy egrets and great egrets, also feed and nest in and adjacent to Rhode Island's coastal wetlands. Great blue herons feed in these wetlands, but nest inland. The U.S. Fish and Wildlife Service (Erwin and Korschgen 1979) has identified nesting colonies of coastal water birds in Rhode Island and other northeastern states. Ospreys also nest in wetlands along the coast.

Southern New England coastal marshes are important feeding and stopover areas for migrating raptors, waterfowl, shorebirds and wading birds. In Rhode Island, intertidal mudflats are principal feeding grounds for migratory shorebirds (e.g., sandpipers, plovers, and yellowlegs), while swallows can often be seen feeding on flying insects over the marshes. The U.S. Fish and Wildlife Service's winter waterfowl survey found an annual average of 9,700 scaup, 3,000 Canada geese, and 2,700 black ducks as well as hundreds of canvasbacks, mallards, mergansers, mute swans, scoters and other waterfowl overwintering in Rhode Island between 1980–1986.

Coastal beaches are used for nesting by piping plover (a Federal threatened species), American oystercatcher, and least tern. Rocky shores are nesting sites for gadwall, double-crested cormorant, roseate tern, and common tern (R. Enser, pers. comm.).

Rhode Island's inland wetlands are used by a variety of birds, including waterfowl, wading birds, rails and songbirds. Among the more typical species are black duck, wood duck, mallard, green-winged teal, Canada goose, mute swan, green-backed heron, great blue heron, least bittern, American bittern, Virginia rail, sora, common moorhen, spotted sandpiper, marsh wren, winter wren, red-winged blackbird, belted kingfisher, tree swallow, northern rough-winged swallow, Acadian flycatcher, willow flycatcher, eastern kingbird, warbling vireo, swamp sparrow, and woodcock. Most of these species are associated with freshwater marshes and open water bodies. Wood duck, Acadian flycatcher, barred owl, northern saw-whet owl, northern waterthrush, Louisiana waterthrush, Canada warbler, and white-throated sparrow nest in forested wetlands. Among the birds breeding in shrub swamps are woodcock and willow flycatcher. Lowry (1984) reported on numerous observations made over a seven-year period in red maple swamps and Atlantic white cedar swamps. Forty-four bird species were seen in the maple swamps, whereas only 25 species were found in cedar swamps. Similar results were reported for southern New Jersey by Wander (1980). Among the birds nesting or assumed to nest in the 30-acre Diamond Bog are mallard, black duck, wood duck, ruffed grouse, Virginia rail, ruby-throated hummingbird, red-winged blackbird, northern oriole, common grackle, common flicker, downy woodpecker, eastern kingbird, great-crested flycatcher, purple finch, American goldfinch, eastern phoebe, tree swallow, blue jay, black-capped chickadee, red-breasted nuthatch, northern waterthrush, common yellowthroat, Canada warbler, American robin, wood thrush, veery, cedar waxwing, black and white warbler, yellow warbler, ovenbird, song sparrow, and swamp sparrow (F. Golet, pers. comm.). (*Note:* Diamond Bog, located in the town of Richmond, is a mosaic of forested, scrub-shrub, and emergent wetlands with some open water.) In a study of eight red maple swamps in western Massachusetts, Swift (1980) found 46 breeding species. The most common breeders included common yellowthroat, veery, Canada warbler, ovenbird, northern waterthrush, and gray catbird. Anderson and Maxfield (1962) studied birdlife in a red maple-Atlantic white cedar swamp in southeastern Massachusetts and found the same species plus ruffed grouse, hairy woodpecker, downy woodpecker, blue jay, black-capped chickadee, American robin, wood thrush, black-and-white warbler, and common grackle.

Wetlands are, therefore, crucial for the existence of many birds, ranging from waterfowl and shorebirds to migratory songbirds. Some spend their entire lives in wetland environments, while others primarily use wetlands for breeding, feeding or resting.

Mammal and Other Wildlife Habitat

Many mammals and other wildlife inhabit Rhode Island wetlands. Muskrats are perhaps the most typical and widespread wetland mammal (Figure 23). Other furbearers inhabiting wetlands include river otter, mink, beaver, raccoon, skunk, red fox, fisher, and weasel. Hardwood swamps are reported to be the favorite habitat of raccoons in Rhode Island (Cronan and Brooks 1968). Beaver populations in the state have been growing since re-introduction in the 1950's. Beaver are most abundant in the Moosup River system in central western Rhode Island (C. Allin, pers. comm.). Smaller mammals also frequent wetlands such as eastern cottontail, New England cottontail, snowshoe hare, meadow vole, boreal red-backed vole, southern bog lemming, water shrew, and meadow jumping mouse, while large mammals may also be observed. White-tailed deer depend on Atlantic white cedar swamps for shelter and food during severe winters, but often use palustrine deciduous forested wetlands and scrub-shrub wetlands for resting and escape cover (Cronan and Brooks 1968; RI DEM, pers. comm.). Another group of mammals—bats—also use wetlands. They can often be seen in considerable numbers feeding over ponds, marshes, and other waterbodies in summer.

Figure 23. The muskrat is the most familiar and widespread wetland mammal in the state.

Besides mammals and birds, other forms of wildlife make their homes in wetlands. Reptiles (i.e., turtles and snakes) and amphibians (i.e., toads, frogs, and salamanders) are important residents. DeGraaf and Rudis (1983) described the non-marine reptiles and amphibians of New England including their habitat and natural history. Turtles are most common in Rhode Island's freshwater marshes and ponds and the more common ones include the eastern painted, spotted, box, stinkpot, wood, and snapping turtles. Common snakes found in and near wetlands include northern water, northern redbelly, eastern garter, eastern ribbon, eastern smooth green, and northern black racer. Among the more common toads and frogs in Rhode Island are Fowler's toad, American toad, northern spring peeper, green frog, bullfrog, wood frog, pickerel frog, and gray tree frog. Less common species include the northern leopard frog (a state special interest species) and the eastern spadefoot (state threatened) (R. Enser, pers. comm.). Adults of the red-spotted newt live in ponds with an abundance of submerged vegetation, while the juveniles are terrestrial. Many salamanders use temporary ponds or wetlands for breeding, although they may spend most of their years in upland or streamside habitats. Nearly all of the approximately 190 species of amphibians in North America are wetland-dependent at least for breeding (Clark 1979). Salamanders common in Rhode Island wetlands include the mudpuppy, spotted, northern dusky, and northern two-lined salamanders, while the four-toed and marbled salamanders are less common and are considered species of concern (R. Enser, pers. comm.).

Rare, Threatened, or Endangered Plants

Currently, the Rhode Island Natural Heritage Program is tracking 261 plant species that are rare, threatened, endangered, or of special interest or concern to the state due to their low numbers (R. Enser, pers. comm.). Of this list, approximately half (132 species) of the plants are considered wetland plants (Table 20). Among the wetland habitats where most of these plants occur are coastal plain pond shores (28 species), salt marshes, estuarine waters, and beaches (15 species), and bogs and fens (15 species).

Environmental Quality Values

Besides providing habitat for fish and wildlife, wetlands play a less conspicuous but essential role in maintaining high environmental quality, especially in aquatic habitats. They do this in a number of ways, including purifying natural waters by removing nutrients, chemical and organic pollutants, and sediment, and producing food which supports aquatic life.

Water Quality Improvement

Wetlands help maintain good water quality or improve degraded waters in several ways: (1) nutrient removal and retention, (2) processing chemical and organic wastes, and (3) reducing sediment load of water. Wetlands are particularly good water filters because of their locations between land and open water (Figure 24). Thus, they can both intercept runoff from land before it reaches the water and help filter nutrients, wastes and sediment from flooding waters. Clean waters are important to humans as well as to aquatic life.

First, wetlands remove nutrients, especially nitrogen and phosphorus, from flooding waters for plant growth and help prevent eutrophication or overenrichment of natural waters. Much of the nutrients are stored in the wetland soil. Freshwater tidal wetlands have proven effective in reducing nutrient and heavy metal loading from surface water runoff from urban areas in the upper Delaware River estuary (Simpson, et al. 1983c). Wetlands in and downstream of urban areas in Rhode Island probably also perform this function. It is, however, possible to overload a wetland and thereby reduce its ability to perform this function. Every wetland has a limited capacity to absorb nutrients and individual wetlands differ in their ability to do so.

Wetlands have been shown to be excellent removers of waste products from water. Sloey and others (1978) summarize the value of freshwater wetlands at removing nitrogen and phosphorus from the water and address management issues. They note that some wetland plants are so efficient at this task that some artificial waste treatment systems are using these plants. For example, the Max Planck Institute of Germany has a patent to create such

Table 20. Plant species of special concern to Rhode Island that occur in wetlands (R. Enser, pers. comm.).

Plant Species	Common Name	State Status[1]
Equisetum fluviatile	Water Horsetail	State Special Interest
Equisetum hyemale	Rough Horsetail	Species of Concern
Lycopodium inundatum var. *robustum*	Northern Bog Clubmoss	State Endangered
Isoetes engelmannii	Engelmann's Quillwort	State Special Interest
Isoetes muricata	Pointed Quillwort	State Special Interest
Isoetes riparia var. *canadensis*	River Quillwort	State Special Interest
Matteuccia struthiopteris	Ostrich Fern	Species of Concern
Larix laricina	American Larch	State Threatened
Picea mariana	Black Spruce	Species of Concern
Sparganium minimum	Small Bur-reed	State Extirpated
Najas guadalupensis	Naiad	State Threatened
Scheuchzeria palustris	Pod Grass	State Endangered
Sagittaria graminea	Grassleaf Arrowhead	State Special Interest
Sagittaria subulata var. *gracillima*	River Arrowhead	State Extirpated
Sagittaria teres	Slender Arrowhead	State Endangered
Panicum philadelphicum	Philadelphia Panic Grass	State Special Interest
Spartina cynosuroides	Salt Reed Grass	State Special Interest
Tripsacum dactyloides	Northern Gamagrass	State Threatened
Zizania aquatica	Wild Rice	Species of Concern
Carex collinsii	Collin's Sedge	State Endangered
Carex exilis	Bog Sedge	State Threatened
Cyperus aristatus	Awned Cyperus	State Extirpated
Eleocharis equisetoides	Horse-tail Spike-rush	State Special Interest
Eleocharis melanocarpa	Black-fruited Spike-rush	State Endangered
Eleocharis tricostata	Three-angle Spike-rush	State Endangered
Eriophorum gracile	Slender Cotton-grass	State Threatened
Eriophorum vaginatum	Hare's Tail	State Endangered
Eriophorum viridicarinatum	Bog Cotton-grass	State Special Interest
Fuirena pumila	Umbrella Grass	State Endangered
Psilocarya scirpoides	Long-beaked Bald Rush	State Endangered
Rhynchospora inundata	Drowned Horned Rush	State Endangered
Rhynchospora macrostachya	Beaked Rush	State Threatened
Rhynchospora torreyana	Torrey's Beaked Rush	State Threatened
Scirpus etuberculatus	Untubercled Bulrush	State Endangered
Scirpus hudsonianus	Cotton Club Rush	State Extirpated
Scirpus longii	Long's Bulrush	State Endangered
Scirpus maritimus var. *fernaldii*	Saltmarsh Bulrush	State Special Interest
Scirpus robustus	Leafy Bulrush	State Special Interest
Scirpus smithii	Smith's Bulrush	State Threatened
Scirpus torreyi	Torrey's Bulrush	State Special Interest
Scleria reticularis	Reticulated Nut-rush	State Threatened
Orontium aquaticum	Golden Club	State Endangered
Xyris montana	Northern Yellow-eyed Grass	State Threatened
Xyris smalliana	Small's Yellow-eyed Grass	Species of Concern
Juncus debilis	Weak Rush	State Special Interest
Alectris farinosa	Colicroot	Species of Concern
Smilacina trifolia	Three-leaved False Solomon's Seal	State Extirpated
Streptopus roseus	Rosy Twisted Stalk	State Threatened
Trillium erectum	Purple Trillium	State Threatened
Lachnanthes caroliniana	Carolina Redroot	State Threatened
Arethusa bulbosa	Swamp Pink	Species of Concern
Calopogon tuberosus	Tuberous Grass Pink	Species of Concern
Corallorhiza trifida	Early Coralroot	State Special Interest
Cypripedium calceolus	Yellow Lady's-slipper	State Threatened
Liparis loeselii	Yellow Twayblade	State Threatened
Malaxis unifolia	Green Adder's Mouth	State Endangered
Platanthera blephariglottis	White-fringed Orchis	State Threatened
Platanthera ciliaris	Yellow-fringed Orchis	State Endangered
Platanthera flava var. *herbiola*	Pale Green Orchis	State Endangered
Platanthera hyperborea	Northern Green Orchis	State Threatened
Platanthera psycodes	Small Puple-fringed Orchid	State Special Interest
Spiranthes lucida	Shining Ladies'-tresses	State Extirpated
Saururus cernuus	Lizard's Tail	State Endangered
Salix pedicellaris	Bog Willow	State Extirpated
Ulmus rubra	Slippery Elm	State Special Interest
Arceuthobium pusillum	Dwarf Mistletoe	State Endangered
Polygonum glaucum	Seabeach Knotweed	State Threatened
Polygonum puritanorum	Pondshore Knotweed	State Endangered
Polygonum setaceum var. *interjectum*	Strigose Knotweed	State Extirpated
Atriplex glabriuscula	Smooth Orache	State Special Interest
Chenopodium leptophyllum	Goosefoot	State Special Interest
Suaeda maritima	Sea-blite	Species of Concern
Amaranthus pumilus	Seabeach Amaranth	State Extirpated
Honkenya peploides	Sea-beach Sandwort	Species of Concern
Anemone riparia	Large Anemone	State Extirpated
Ranunculus aquatilis	White Water Crowfoot	State Extirpated
Ranunculus cymbalaria	Seaside Buttercup	State Extirpated
Ranunculus flabellaris	Yellow Water Crowfoot	Species of Concern

Table 20. (*Continued*)

Plant Species	Common Name	State Status[1]
Draba reptans	Carolina Whitlow-Grass	State Extirpated
Drosera filiformis	Thread-leaved Sundew	State Endangered
Podostemum ceratophyllum	Riverweed	State Extirpated
Parnassia glauca	Grass-of-Parnassus	State Extirpated
Saxifraga pensylvanica	Swamp Saxifrage	State Threatened
Dalibarda repens	Dewdrop	State Endangered
Crotalaria sagittalis	Rattlebox	State Threatened
Polygala cruciata	Cross-leaved Milkwort	State Threatened
Hypericum adpressum	Creeping St. John's-wort	State Threatened
Hypericum ellipticum	Pale St. John's-wort	State Special Interest
Viola incognita	Large-leaf White Violet	State Special Interest
Elatine americana	American Waterwort	State Special Interest
Rotala ramosior	Toothcup	State Endangered
Circaea alpina	Small Enchanter's Nightshade	Species of Concern
Epilobium palustre	Marsh Willow-herb	State Special Interest
Ludwigia sphaerocarpa	Round-fruited False Loosestrife	State Endangered
Myriophyllum alterniflorum	Alternate-flowered Water-milfoil	State Extirpated
Myriophyllum pinnatum	Pinnate Water-milfoil	State Extirpated
Angelica atropurpurea	Large Angelica	State Extirpated
Hydrocotyle verticillata	Saltpond Pennywort	State Endangered
Ligusticum scothicum	Scotch Lovage	State Threatened
Ptilimnium capillaceum	Mock Bishop's Weed	State Special Interest
Andromeda polifolia	Bog Rosemary	State Endangered
Gaultheria hispidula	Creeping Snowberry	State Special Interest
Gaylussacia dumosa var. *bigeloviana*	Dwarf Huckleberry	Species of Concern
Kalmia polifolia	Pale Laurel	State Endangered
Leucothoe racemosa var. *projecta*	Projecting Fetter-bush	Species of Concern
Rhododendron periclymenoides	Pinxter-flower	State Extirpated
Glaux maritima	Sea Milkwort	State Extirpated
Hottonia inflata	Featherfoil	State Special Interest
Fraxinus nigra	Black Ash	Species of Concern
Gentiana andrewsii	Closed Gentian	State Extirpated
Gentiana clausa	Bottle Gentian	State Special Interest
Gentianopsis crinita	Fringed Gentian	State Threatened
Sabatia kennedyana	Plymouth Gentian	State Endangered
Sabatia stellaris	Sea Pink	State Threatened
Physostegia virginiana	False Dragon-head	State Special Interest
Stachys hyssopifolia	Hyssop-leaf Hedge-nettle	State Endangered
Agalinis maritima	Seaside Gerardia	Species of Concern
Limosella australis	Mudwort	Species of Concern
Utricularia biflora	Two-flower Bladderwort	State Threatened
Utricularia geminiscapa	Paired Bladderwort	State Special Interest
Utricularia gibba	Humped Bladderwort	State Special Interest
Utricularia intermedia	Flatleaf Bladderwort	State Special Interest
Utricularia minor	Small Bladderwort	State Extirpated
Utricularia resupinata	Reversed Bladderwort	State Threatened
Utricularia subulata	Zigzag Bladderwort	State Threatened
Viburnun nudum	Swamp-haw	State Threatened
Lobelia dortmanna	Water Lobelia	Species of Concern
Bidens connata	Swamp Beggar-ticks	State Special Interest
Bidens coronata	Tickseed Sunflower	State Special Interest
Coreopsis rosea	Pink Tickseed	State Threatened
Eupatorium leucolepis var. *novae-angliae*	New England Boneset	State Endangered
Sclerolepis uniflora	Sclerolepis	State Endangered

[1] Definitions of State Status:

"State Endangered" are native species in imminent danger of extirpation from Rhode Island; these species meet one or more of the following criteria:
 1. A species currently listed, or proposed by the U.S. Fish and Wildlife Service as Federally endangered or threatened.
 2. A species with 1 or 2 known or estimated total occurrences in the state.
 3. A species apparently globally rare or threatened, and estimated to occur at approximately 100 or fewer occurrences range-wide.

"State Threatened" are native species which are likely to become state endangered in the future if current trends in habitat loss or other detrimental factors remain unchanged; these species meet one or more of the following criteria:
 1. A species with 3 to 5 known or estimated occurrences in the state.
 2. A species with more than 5 known or estimated occurrences in the state, but especially vulnerable to habitat loss.

"State Special Interest" are native species not considered to be State Endangered or State Threatened at the present time, but occur in 6 to 10 sites in the state.

"Species of Concern" are native species which do not apply under the above categories but are additionally listed by the Natural Heritage Program due to various factors of rarity and/or vulnerability.

"State Extirpated" are native species which have been documented as occurring in the state but for which current occurrences are unknown. When known, the last documentation of occurrence is included. If an occurrence is located for a State Extirpated species, that species would automatically be listed in the State Endangered category.

Figure 24. Wetlands are important for water quality improvement as well as flood water storage. Their location between the upland and the water facilitates these functions.

systems, where a bulrush (*Scirpus lacustris*) is the primary waste removal agent. Numerous scientists have proposed that certain types of wetlands be used to process domestic wastes and some wetlands are already used for this purpose (Sloey, *et al.* 1978; Carter, *et al.* 1979; Kadlec 1979). It must, however, be recognized that individual wetlands have a finite capacity for natural assimilation of excess nutrients and research is needed to determine this threshold (Good 1982). In the meantime, it may be prudent to use artificial wetlands for treatment of secondary wastes and then run the tertiary products into a natural wetland, rather than having natural wetlands process the entire wasteload. Godfrey and others (1985) discuss ecological considerations of using wetlands to treat municipal wastewaters.

Perhaps the best known example of the importance of wetlands for water quality improvement is Tinicum Marsh (Grant and Patrick 1970). Tinicum Marsh is a 512-acre freshwater tidal marsh lying just south of Philadelphia, Pennsylvania. Three sewage treatment plants discharge treated sewage into marsh waters. On a daily basis, it was shown that this marsh removes from flooding waters: 7.7 tons of biological oxygen demand, 4.9 tons of phosphorus, 4.3 tons of ammonia, and 138 pounds of nitrate. In addition, Tinicum Marsh adds 20 tons of oxygen to the water each day.

Swamps also have the capacity for removing water pollutants. Bottomland forested wetlands along the Alcovy River in Georgia have been shown to filter impurities from flooding waters. Human and chicken wastes grossly pollute the river upstream, but after passing through less than 3 miles of swamp, the river's water quality is significantly improved. The value of the 2,300-acre Alcovy River Swamp for water pollution control was estimated at $1 million per year (Wharton 1970). In New Jersey, Durand and Zimmer (1982) have demonstrated the capacity of Pine Barrens wetlands to assimilate excess nutrients from adjacent agricultural land and upland development. Rhode Island's wetlands undoubtedly function similarly to these wetlands.

Wetlands also play a valuable role in reducing turbidity of flooding waters. This is especially important for aquatic life and for reducing siltation of ports, harbors, rivers and reservoirs. Removal of sediment load is also valuable because sediments often transport adsorbed nutrients,

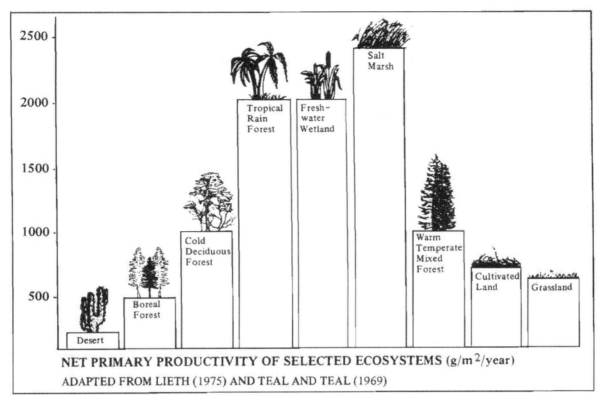

Figure 25. Relative productivity of wetland ecosystems in relation to other ecosystems (redrawn from Newton 1981). Salt marshes and freshwater marshes are among the world's most productive systems.

pesticides, heavy metals and other toxins which pollute our Nation's waters (Boto and Patrick 1979). Depressional wetlands should retain all of the sediment entering them (Novitzki 1978). In Wisconsin, watersheds with 40 percent coverage by lakes and wetlands had 90 percent less sediments in water than watersheds with no lakes or wetlands (Hindall 1975). Creekbanks of salt marshes typically support more productive vegetation than the marsh interior. Deposition of silt is accentuated at the water-marsh interface, where vegetation slows the velocity of water, causing sediment to drop out of solution. In addition to improving water quality, this process adds nutrients to the creekside marsh which leads to higher plant density and plant productivity (DeLaune, et al. 1978).

The U.S. Army Corps of Engineers has investigated the use of marsh vegetation to lower turbidity of dredged disposal runoff and to remove contaminants. In a 50-acre dredged material disposal impoundment near Georgetown, South Carolina, after passing through about 2,000 feet of marsh vegetation, the effluent turbidity was similar to that of the adjacent river (Lee, et al. 1976). Wetlands have also been proven to be good filters of nutrients and heavy metal loads in dredged disposal effluents (Windom 1977).

Recently, the ability of wetlands to retain heavy metals has been reported (Banus, et al. 1974; Mudroch and Capobianca 1978; Simpson, et al. 1983c). Wetland soils have been regarded as primary sinks for heavy metals, while wetland plants may play a more limited role. Waters flowing through urban areas often have heavy concentrations of heavy metals (e.g., cadmium, chromium, copper, nickel, lead, and zinc). The ability of freshwater tidal wetlands along the Delaware River in New Jersey to sequester and hold heavy metals has been documented (Good, et al. 1975; Whigham and Simpson 1976; Simpson et al. 1983a, 1983b, 1983c). Wetlands along heavily industrialized rivers in Rhode Island probably are retaining various heavy metals also. Additional study is needed to better understand retention mechanisms and capacities in wetlands.

Aquatic Productivity

Wetlands are among the most productive ecosystems in the world and they may be the highest, rivaling our best cornfields (Figure 25). Wetland plants are particularly efficient converters of solar energy. Through photosynthesis, plants convert sunlight into plant material or biomass and produce oxygen as a by-product. Other mate-

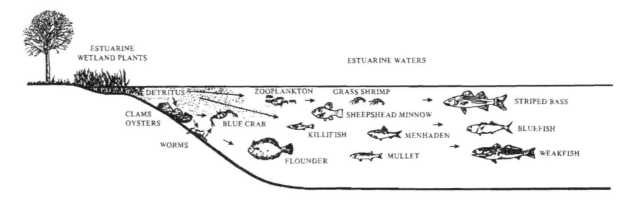

Figure 26. Simplified food pathways from estuarine wetland vegetation to commercially and recreationally important fishes and shellfishes.

rials, such as organic matter, nutrients, heavy metals, and sediment, also are captured by wetlands and either stored in the sediment or converted to biomass (Simpson, *et al.* 1983a). This biomass serves as food for a multitude of animals, both aquatic and terrestrial. For example, many waterfowl depend heavily on seeds of marsh plants, while muskrats eat cattail tubers and young shoots. Surprisingly, one of the favorite winter foods of the eastern cottontail is the tender new growth of red maples (Cronan and Brooks 1968).

Although direct grazing of wetland plants may be considerable in freshwater marshes, their major food value to most aquatic organisms is reached upon death when plants break down to form "detritus." This detritus forms the base of an aquatic food web that supports higher consumers, e.g., commercial fishes. This relationship is especially well-documented for coastal areas. Animals like zooplankton, shrimp, snails, clams, worms, killifish, and mullet eat detritus or graze upon the bacteria, fungi, diatoms and protozoa growing on its surfaces (Crow and Macdonald 1979; de la Cruz 1979). Forage fishes (e.g., anchovies, sticklebacks, killifishes, and silversides) and grass shrimp are the primary food for commercial and recreational fishes, including bluefish, flounder, weakfish, and white perch (Sugihara, *et al.* 1979). A simplified food web for estuaries in the Northeast is presented as Figure 26. Thus, wetlands can be regarded as the farmlands of the aquatic environment where great volumes of food are produced annually. The majority of non-marine aquatic animals also depend, either directly or indirectly, on this food source.

Socio-economic Values

The more tangible benefits of wetlands to society may be considered socio-economic values and they include flood and storm damage protection, erosion control, water supply and ground-water recharge, harvest of natural products, livestock grazing and recreation. Since these values provide either dollar savings or financial profit, they are more easily understood by most people.

Flood and Storm Damage Protection

In their natural condition, wetlands serve to temporarily store flood waters, thereby protecting downstream property owners from flood damage. After all, such flooding has been the driving force in creating these wetlands to begin with. This flood storage function also helps to slow the velocity of water and lower wave heights, reducing the water's erosive potential. Rather than having all flood waters flowing rapidly downstream and destroying private property and crops, wetlands slow the flow of water, store it temporarily and slowly release stored waters downstream (Figure 27). Wetlands, thereby, help reduce the peak flood heights as well as delay the flood crest. This becomes increasingly important in urban areas, where development has increased the rate and volume of surface water runoff and the potential for flood damage (Figure 28).

In 1975, 107 people were killed by flood waters in the U.S. and potential property damage for the year was estimated to be $3.4 billion (U.S. Water Resources Council 1978). Almost half of all flood damage was suffered by farmers as crops and livestock were destroyed and productive land was covered by water or lost to erosion. Approximately 134 million acres of the conterminous U.S. have severe flooding problems (Figure 29). Of this, 2.8 million acres are urban land and 92.8 million acres are agricultural land (U.S. Water Resources Council 1977). Many of these flooded farmlands are wetlands. Although regulations and ordinances required by the Federal Insurance Administration reduce flood losses from urban land, agricultural losses are expected to remain at present levels

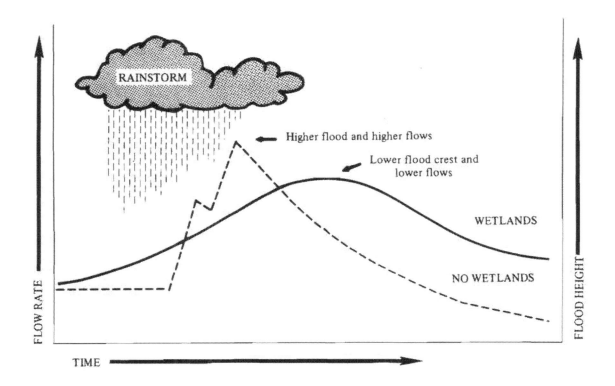

Figure 27. Wetlands help reduce flood crests and slow flow rates after rainstorms (adapted from Kusler 1983).

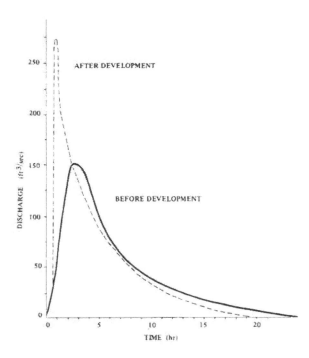

Figure 28. Urban development increases peak discharge in rivers. Comparisons of hydrographs for a watershed before and after development (redrawn from Fusillo 1981).

or increase as more wetland is put into crop production. Protection of wetlands is, therefore, an important means to minimizing flood damages in the future.

The U.S. Army Corps of Engineers have recognized the value of wetlands for flood storage in Massachusetts. In the early 1970's, they considered various alternatives to providing flood protection in the lower Charles River watershed near Boston, including: (1) a 55,000 acre-foot reservoir, (2) extensive walls and dikes, and (3) perpetual protection of 8,500 acres of wetland (U.S. Army Corps of Engineers 1976). If 40 percent of the Charles River wetlands were destroyed, flood damages would have increased by at least $3 million annually. Loss of all basin wetlands would cause an average annual flood damage cost of $17 million (Thibodeau and Ostro 1981). The Corps concluded that wetlands protection—"Natural Valley Storage"—was the least-cost solution to future flooding problems. In 1983, they completed acquisition of approximately 8,500 acres of Charles River wetlands for flood protection.

This protective value of wetlands has also been reported for other areas. Undeveloped floodplain wetlands in New Jersey protect against flood damages (Robichaud and Buell 1973). In the Passaic River watershed, annual property losses to flooding approached $50 million in

Figure 29. Wetland destruction accelerates flood damages.

1978 and the Corps of Engineers is considering wetland acquisition as an option to prevent flood damages from escalating in the future (U.S. Army Corps of Engineers 1979). A Wisconsin study projected that floods may be lowered as much as 80 percent in watersheds with many wetlands compared with similar basins with few or no wetlands (Novitzki 1978). Pothole wetlands in the Devils Lake basin of North Dakota store nearly 75 percent of the total runoff (Ludden, et al. 1983).

Rhode Island's wetlands also serve as temporary storage basins for retaining flood waters, thereby reducing potential flood damages. The 3,000-acre Great Swamp, Chapman Swamp and numerous other wetlands provide great flood storage for the Pawcatuck River in Washington County and without these wetlands flooding of downstream uplands would be enormous. The Pawtuxet River system, in marked contrast, has fewer wetlands (many wetlands were filled) and less flood storage area. Consequently, Warwick and Cranston experience serious flooding problems. Annual flood losses in 1978 for the Pawtuxet River basin were about $1.5 million. Corps of Engineers projections for 1990 suggest that increased urbanization in the basin would raise flood losses to $3.6 million for a 20-year flood and $5.5 million for a 50-year flood (F. Golet, pers. comm.).

Shoreline Erosion Control

Located between watercourses and uplands, wetlands help protect uplands from erosion. Wetland vegetation can reduce shoreline erosion in several ways, including: (1) increasing durability of the sediment through binding with its roots, (2) dampening waves through friction, and (3) reducing current velocity through friction (Dean 1979). This process also helps reduce turbidity and thereby helps improve water quality.

Obviously, trees are good stabilizers of river banks. Their roots bind the soil, making it more resistant to erosion, while their trunks and branches slow the flow of flooding waters and dampen wave heights. The banks of some rivers have not been eroded for 100 to 200 years due to the presence of trees (Leopold and Wolman 1957; Wolman and Leopold 1957; Sigafoos 1964). Among the freshwater grass and grass-like plants, common reed (*Phragmites australis*) and bulrushes (*Scirpus* spp.) have been regarded as the best at withstanding wave and current action (Kadlec and Wentz 1974; Seibert 1968). Common three-square (*Scirpus pungens*) often forms fringing marshes along the margins of many Rhode Island lakes and ponds. Along the coast, salt marshes of smooth cordgrass (*Spartina alterniflora*) are considered important shoreline stabilizers because of their wave dampening effect (Knudson, et al. 1982). While most wetland plants need calm or sheltered water for establishment, they will effectively control erosion once established (Kadlec and Wentz 1974; Garbisch 1977). Wetland vegetation has been successfully planted to reduce erosion along U.S. waters. Willows (*Salix* spp.), alders (*Alnus* spp.), ashes (*Fraxinus* spp.), cottonwoods and poplars (*Populus* spp.), maples (*Acer* spp.), and elms (*Ulmus* spp.) are particularly good stabilizers (Allen 1979). Successful emergent plants include reed canary grass (*Phalaris arundinacea*), common reed, cattails (*Typha* spp.), and bulrushes in freshwater areas (Hoffman 1977) and smooth cordgrass along the coast (Woodhouse, et al. 1976).

Water Supply

Most wetlands are areas of ground-water discharge and their underlying aquifers may provide sufficient quantities of water for public use. In neighboring Massachusetts, 40 percent to 50 percent of the wetlands may indicate the location of productive underground aquifers—potential sources of drinking water. At least 60 municipalities in the state have public wells in or very near wetlands (Motts and Heeley 1973). Prairie pothole wetlands store water which is important for wildlife and may be used for irrigation and livestock watering by farmers during droughts (Leitch

Figure 30. Cows often graze in wet meadows.

1981). These situations may hold true for Rhode Island and other states. Wetland protection and ground-water pollution control could be instrumental in helping to solve current and future water supply problems.

Ground-water Recharge

Ground-water recharge potential of wetlands varies according to numerous factors, including wetland type, geographic location, season, soil type, water table location and precipitation. In general, most researchers believe that most wetlands do not serve as significant ground-water recharge sites (Carter, *et al.* 1979). A few studies, however, have shown that certain wetland types may help recharge ground-water supplies by adding water to the underlying aquifer or water table. Shrub wetlands in the Pine Barrens may contribute to ground-water recharge (Ballard 1979). Depressional wetlands, like cypress domes in Florida and prairie potholes in the Dakotas, may also contribute to ground-water recharge (Odum, *et al.* 1975; Stewart and Kantrud 1972).

Floodplain wetlands also may do this through bank water storage (Mundorff 1950; Klopatek 1978). In urban areas where municipal wells pump water from streams and adjacent wetlands, "induced infiltration" may draw in surface water from wetlands into public wells. This type of human-induced recharge has been observed in Burlington, Massachusetts (Mulica 1977). These studies and others suggest that certain wetlands do help recharge ground-water and that additional research is needed to better assess the role of different types of wetlands in performing this function.

Harvest of Natural Products

A variety of natural products are produced by wetlands including timber, fish and shellfish, wildlife, peat moss, cranberries, blueberries, and wild rice. Wetland grasses are hayed in many places for winter livestock feed. During other seasons, livestock graze directly in numerous New England wetlands (Figure 30).

In the 49 continental states, an estimated 82 million acres of commercial forested wetlands exist (Johnson 1979). These forests provide timber for such uses as home construction, furniture, newspapers and firewood. Most of these forests lie east of the Rockies, where oak, gum, cypress, elm, ash and cottonwood are most important. The standing value of southern wetland forests is $8 billion. These southern forests have been harvested for over 200 years without noticeable degradation, thus they can be expected to produce timber for many years to come, unless converted to other uses. Rhode Island's forested wetlands provide timber for fuelwood and building construction. Braiewa (1983) reported on the biomass and fuelwood production of red maple stands in the state.

Many wetland-dependent fishes and wildlife are also utilized by society. Commercial fishermen and trappers make a living from these resources. From 1956 to 1975, about 60 percent of the U.S. commercial landings were fishes and shellfishes that depend on wetlands (Peters, *et al.* 1979). Nationally, major commercial species associated with wetlands are menhaden, salmon, shrimp, blue crab and alewife from coastal waters and catfish, carp and buffalo from inland areas. In Rhode Island, the 1985 commercial harvest of wetland-dependent coastal fishes (i.e., flounders, bluefish, weakfish, striped bass, shad, and white perch) had a value of $3.25 million, while the hard-shell clam or quahog harvest alone was valued at more than $14 million according to National Marine Fisheries Service commercial catch and value data. The fisheries value of Rhode Island's coastal ponds is discussed by Lee (1980). Recreational fishing and shellfishing are important activities for many Rhode Island residents.

Nationally, furs from beaver, muskrat, mink, nutria, and otter yielded roughly $35.5 million in 1976 (Demms and Pursley 1978). Louisiana is the largest fur-producing state and nearly all furs come from wetland animals. In Rhode Island, muskrat harvest was valued at near $60,000 in 1980 and only about $6,500 in 1988 due to declining pelt prices (L. Suprock and M. Lapisky, pers. comm.). Currently, muskrats are an under-harvested resource.

Recreation and Aesthetics

Many recreational activities take place in and around wetlands. Hunting and fishing are popular sports. Waterfowl hunting is a major activity in wetlands, but big game hunting is also important locally. In 1980, 5.3 million people spent $638 million on hunting waterfowl and other migratory birds (U.S. Department of the Interior and Department of Commerce 1982). Moreover, nearly all freshwater fishing is dependent on wetlands. In 1975 alone, sportfishermen spent $13.1 billion to catch wetland-dependent fishes in the U.S. (Peters, et al. 1979). Fishing was reported to be the second most popular leisure sport in America in a 1985 Gallup Poll (Sport Fishing Institute 1986). Fishing was the top activity for adult men with 44 percent participating. Since 1977, there has been a steady increase in the percent of Americans fishing.

Other recreation in wetlands is largely non-consumptive and involves activities like hiking, nature observation and photography, and canoeing and other boating. Many people simply enjoy the beauty and sounds of nature and spend their leisure time walking or boating in or near wetlands and observing plant and animal life. This aesthetic value is extremely difficult to place a dollar value upon, although people spend a great deal of money traveling to places to enjoy the scenery and to take pictures of these scenes and plant and animal life. In 1980, 28.8 million people (17 percent of the U.S. population) took special trips to observe, photograph or feed wildlife. Moreover, about 47 percent of all Americans showed an active interest in wildlife around their home (U.S. Department of the Interior and Department of Commerce 1982).

Summary

Marshes, swamps and other wetlands are assets to society in their natural state. They provide numerous products for human use and consumption, protect private property and provide recreational and aesthetic appreciation opportunities. Wetlands may also have other values yet unknown to society. For example, a microorganism from Pine Barrens swamps of southern New Jersey has been recently discovered to have great value to the drug industry. In searching for a new source of antibiotics, the Squibb Institute examined soils from around the world and found that only one contained microbes suitable for producing a new family of antibiotics. From a Pine Barrens swamp microorganism, scientists at the Squibb Institute have developed a new line of antibiotics which will be used to cure diseases not affected by present antibiotics (Moore 1981). This represents a significant medical discovery. If these wetlands were destroyed or grossly polluted, this discovery might not have been possible.

Destruction or alteration of wetlands eliminates or minimizes their values. Drainage of wetlands, for example, eliminates all the beneficial effects of the wetlands on water quality and directly contributes to flooding problems (Lee, et al. 1975). While the wetland landowner can derive financial profit from some of the values mentioned, the general public receives the vast majority of wetland benefits through flood and storm damage control, erosion control, water quality improvement and fish and wildlife resources. It is, therefore, in the public's best interest to protect wetlands to preserve these values for themselves and future generations. Since over half of the Nation's original wetlands have already been destroyed, the remaining wetlands are even more valuable as public resources.

References

Allen, H.H. 1979. Role of wetland plants in erosion control of riparian shorelines. In: P.E. Greeson, et al. Wetland Functions and Values: The State of Our Understanding. Amer. Water Resources Assoc. pp. 403–414.

Anderson, K.S. and H.K. Maxfield. 1962. Sampling passerine birds in a wooded swamp in southeastern Massachusetts. Wilson Bull. 74(4): 381–385.

Ballard, J.T. 1979. Fluxes of water and energy through the Pine Barrens ecosystems. In: R.T.T. Forman (editor). Pine Barrens: Ecosystem and Landscape. Academic Press, Inc., New York. pp. 133–146.

Banus, M., I. Valiela, and J.M. Teal. 1974. Export of lead from salt marshes. Mar. Poll. Bull. 5: 6–9.

Boto, K.G. and W.H. Patrick, Jr. 1979. Role of wetlands in the removal of suspended sediments. In: P.E. Greeson, et al. Wetland Functions and Values: The State of Our Understanding. Amer. Water Resources Assoc. pp. 479–489.

Braiewa, M.A. 1983. Biomass and Fuelwood Production of Red Maple (Acer rubrum) stands in Rhode Island. M.S. theseis, University of Rhode Island, Kingston. 134 pp.

Carter, V., M.S. Bedinger, R.P. Novitski and W.O. Wilen. 1979. Water resources and wetlands. In: P.E. Greeson, et al. Wetland Functions and Values: The State of Our Understanding. Amer. Water Resources Assoc. pp. 344–376.

Clark, J.E. 1979. Fresh water wetlands: habitats for aquatic invertebrates, amphibians, reptiles, and fish. *In:* P. E. Greeson, *et al.* Wetland Functions and Values: The State of Our Understanding. Amer. Water Resources Assoc. pp. 330–343.

Cronan, J.M. and A. Brooks. 1968. The Mammals of Rhode Island. Rhode Island Dept. of Natural Resources. Wildlife Pamphlet No. 6. 133 pp.

Crow, J.H. and K.B. MacDonald. 1979. Wetland values: secondary production. *In:* P.E. Greeson, *et al.* Wetland Functions and Values: The State of Our Understanding. Amer. Water Resources Assoc. pp. 146–161.

Davenport, C.B. 1903. Animal ecology of the Cold Spring Harbor sand spit. Decennial Pub. Univ. of Chicago 11: 157–176.

Dean, R.G. 1979. Effects of vegetation on shoreline erosional processes. *In:* P.E. Greeson, *et al.* Wetland Functions and Values: The State of Our Understanding. Amer. Water Resources Assoc. pp. 415–426.

DeGraaf, R.M. and D.D. Rudis. 1983. Amphibians and Reptiles of New England: Habitats and Natural History. University of Massachusetts Press, Amherst. 85 pp.

de la Cruz, A.A. 1979. Production and transport of detritus in wetlands. *In:* P.E. Greeson, *et al.* Wetlands Functions and Values: The State of Our Understanding. Amer. Water Resources Assoc. pp. 162–174.

DeLaune, R.D., W.H. Patrick, Jr. and R.J. Buresk. 1978. Sedimentation rates determined by 137Cs dating in a rapidly accreting salt marsh. Nature 275: 532–533.

Demms, E.F., Jr. and D. Pursley (editors). 1978. North American Furbearers: Their Management, Research and Harvest Status in 1976. International Assoc. of Fish and Wildlife Agencies. 157 pp.

Durand, J.B. and B. Zimmer. 1982. Pinelands Surface Water Quality. Part I. Rutgers Univ., Center for Coastal and Environmental Studies, New Brunswick, NJ. 196 pp.

Erwin, R.M. and C.E. Korschgen. 1979. Coastal Waterbird Colonies: Maine to Virginia, 1977. U.S. Fish and Wildlife Service. FWS/OBS-79/08. 647 pp. + appendices.

Fusillo, T.V. 1981. Impact of Suburban Residential Development on Water Resources in the Area of Winslow Township, Camden County, New Jersey. U.S. Geol. Survey, Water Resources Div., Trenton, N.J. Water Resources Investigations 8–27. 28 pp.

Garbisch, E.W., Jr. 1977. Marsh development for soil erosion. *In:* Proc. of the Workshop on the Role of Vegetation in Stabilization of the Great Lakes Shoreline. Great Lakes Basin Commission, Ann Arbor, MI. pp. 77–94.

Godfrey, P.J., E.R. Kaynor, S. Pelczarski, and J. Benforado (editors). 1985. Ecological Considerations in Wetlands Treatment of Municipal Wastewaters. Van Nostrand Reinhold Co., New York. 474 pp.

Good, R.E. (editor). 1982. Ecological Solutions to Environmental Management Concerns in the Pinelands National Reserve. Proceedings of a conference, April 18–22, 1982. Rutgers Univ., Div. of Pinelands Research, New Brunswick, NJ. 47 pp.

Good, R.E., J. Limb, E. Lyszczek, M. Miernik, C. Ogrosky, N. Psuty, J. Ryan, and F. Stickels. 1978. Analysis and Delineation of Submerged Vegetation of Coastal New Jersey: A Case Study of Little Egg Harbor. Rutgers Univ., Center for Coastal and Environmental Studies, New Brunswick, NJ. 58 pp.

Good, R.E., R.W. Hastings, and R.E. Denmark. 1975. An Environmental Assessment of Wetlands: A Case Study of Woodbury Creek and Associated Marshes. Rutgers Univ., Mar. Sci. Ctr., New Brunswick, NJ. Tech. Rept. 75-2. 49 pp.

Grant, R.R., Jr. and R. Patrick. 1970. Tinicum Marsh as a water purifier. *In:* Two Studies of Tinicum Marsh. The Conservation Foundation, Washington, D.C. pp. 105–123.

Greeson, P.B., J.R. Clark and J.E. Clark (editors). 1979. Wetland Functions and Values: The State of Our Understanding. Proc. of the National Symposium on Wetlands. November 7–10, 1978. Amer. Water Resources Assoc., Minneapolis, Minnesota. 674 pp.

Guthrie, R.C. and J.A. Stolgitis. 1977. Fisheries Investigations and Management in Rhode Island Lakes and Ponds. R.I. Dept. of Nat. Res., Div. of Fish and Wildlife. Fisheries Rept. No. 3. 256 pp.

Hindall, S.M. 1975. Measurements and Prediction of Sediment Yields in Wisconsin Streams. U.S. Geological Survey Water Resources Investigations 54–75. 27 pp.

Hoffman, G.R. 1977. Artificial establishment of vegetation and effects of fertilizer along shorelines of Lake Oahe and Sakakawea mainstream Missouri River reservoirs. *In:* Proc. Workshop on the Role of Vegetation in Stabilization of the Great Lakes Shoreline. Great Lakes Basin Commission, Ann Arbor, MI. pp. 95–109.

Johnson, R.L. 1979. Timber harvests from wetlands. *In:* P.E. Greeson, *et al.* Wetland Functions and Values: The State of Our Understanding. Amer. Water Resources Assoc. pp. 598–605.

Kadlec, J.A. and W.A. Wentz. 1974. State-of-the-art Survey and Evaluation of Marsh Plant Establishment Techniques: Induced and Natural. Vol. I: Report of Research. Tech. Rept. D-74-9. U.S. Army Engineers Waterways Expt. Stat., Vicksburg, MS.

Kadlec, R.H. 1979. Wetlands for tertiary treatment. *In:* P.E. Greeson, *et al.* Wetland Functions and Values: The State of Our Understanding. Amer. Water Resources Assoc. pp. 490–504.

Klopatek, J.M. 1978. Nutrient dynamics of freshwater riverine marshes and the role of emergent macrophytes. *In:* R.E. Good, D.F. Whigham, and R.L. Simpson (editors). 1978. Freshwater Wetlands. Ecological Processes and Management Potential. Academic Press Inc., New York. pp. 195–216.

Knudson, P.L., R.A. Brocchu, W.N. Seelig and M. Inskeep. 1982. Wave dampening in *Spartina alterniflora* marshes. Wetlands (Journal of the Society of Wetland Scientists) 2: 87–104.

Kusler, J.A. 1983. Our National Wetland Heritage. A Protection Guidebook. Environmental Law Institute, Washington, DC. 167 pp.

Lee, C.R., R.E. Hoeppel, P.G. Hunt and C.A. Carlsong. 1976. Feasibility of the Functional Use of Vegetation to Filter, Dewater, and Remove Contaminants from Dredged Material. Tech. Rept. D-76-4. U.S. Army Engineers, Waterways Expt. Sta., Vicksburg, MS.

Lee, G.F., E. Bentley, and R. Amundson. 1975. Effects of marshes on water quality. *In:* A.D. Hasler (editor). Coupling of Land and Water Systems. Springer-Verlag, New York. pp. 105–127.

Lee, V. 1980. An Elusive Compromise: Rhode Island Coastal Ponds and Their People. University of Rhode Island, Coastal Res. Ctr., Narragansett. Marine Tech. Rept. 73. 82 pp.

Leitch, J.A. 1981. Wetland Hydrology: State-of-the-art and Annotated Bibliography. Agric. Expt. Stat., North Dakota State Univ., Fargo, ND. Res. Rept. 82. 16 pp.

Leopold, L.B. and M.G. Wolman. 1957. River Channel Patterns—Braided, Meandering, and Straight. U.S. Geol. Survey Prof. Paper 282-B.

Lowry, D. 1984. Water Regimes and Vegetation of Rhode Island Forested Wetlands. M.S. thesis, Univ. of Rhode Island, Kingston. 174 pp.

Ludden, A.P., D.L. Frink, and D.H. Johnson. 1983. Water storage capacity of natural wetland depressions in the Devils Lake Basin of North Dakota. J. Soil and Water Cons. 38(1): 45–48.

McHugh, J.L. 1966. Management of Estuarine Fishes. Amer. Fish Soc., Spec. Pub. No. 3: 133–154.

Moore, M. 1981. Pineland germ yields new antibiotic. Sunday Press (September 6, 1981), Atlantic City, NJ.

Motts, W.S. and R.W. Heeley. 1973. Wetlands and groundwater. *In:* J.S. Larson (editor). A Guide to Important Characteristics and Values of Freshwater Wetlands in the Northeast. University of Massachusetts, Water Resources Research Center. Pub. No. 31. pp. 5–8.

Mudroch, A. and J. Capobianco. 1978. Study of selected metals in marshes on Lake St. Clair, Ontario. Archives Hydrobiologic. 84: 87–108.

Mulica, W.S. 1977. Wetlands and Municipal Ground Water Resource Protection. *In:* R.B. Pojasek (editor). Drinking Water Quality Enhancement Through Source Protection Ann Arbor Science Publishers, Michigan. pp. 297–316.

Mundorff, M.J. 1950. Floodplain Deposits of North Carolina Piedmont and Mountain Streams as a Possible Source of Groundwater Supply. N.C. Div. Mineral Res. Bull. 59.

Newton, R.B. 1981. New England Wetlands: A Primer. University of Massachusetts, Amherst. M.S. thesis. 84 pp.

Novitzki, R.P. 1978. Hydrology of the Nevin Wetland Near Madison, Wisconsin. U.S. Geological Survey, Water Resources Investigations 78–48. 25 pp.

Peters, D.S., D.W. Ahrenholz, and T.R. Rice. 1979. Harvest and value of wetlands associated fish and shellfish. *In:* Greeson, *et al.* Wetland Functions and Values: The State of Our Understanding. Amer. Water Resources Assoc. pp. 606–617.

Reinert, S.E., F.C. Golet, and W.R. DeRagon. 1981. Avian use of ditched and unditched salt marshes in southeastern New England: a preliminary report. *In:* Proceedings from the 27th Annual Meeting of the Northeastern Mosquito Control Association (November 2–4, 1981, Newport, RI). pp. 1–20.

Robichaud, B. and M.F. Buell. 1973. Vegetation of New Jersey: A Study of Landscape Diversity. Rutgers Univ. Press. New Brunswick, NJ. 340 pp.

Seibert, P. 1968. Importance of natural vegetation for the protection of the banks of streams, rivers, and canals. *In:* Nature and Environment Series (Vol. Freshwater), Council of Europe. pp. 35–67.

Sigafoos, R.S. 1964. Botanical Evidence of Floods and Floodplain Deposition, Vegetation, and Hydrologic Phenomena. U.S. Geol. Survey Prof. Paper 485-A.

Simpson, R.L., R.E. Good, B.J. Dubinski, J.J. Pasquale and K.R. Philipp. 1983a. Fluxes of Heavy Metals in Delaware River Freshwater Tidal Wetlands. Rutgers University, Center for Coastal and Environmental Studies, New Brunswick, NJ. 79 pp.

Simpson, R.L., R.E. Good, M.A. Leck, and D.F. Whigham. 1983b. The ecology of freshwater tidal wetlands. BioScience 33(4): 255–259.

Simpson, R.L., R.E. Good, R. Walker, and B.R. Frasco. 1983c. The role of Delaware River freshwater tidal wetlands in the retention of nutrients and heavy metals. J. Environ. Qual. 12(1): 41–48.

Sloey, W.E., R.L. Spangler, and C.W. Fetter, Jr. 1978. Management of freshwater wetlands for nutrient assimilation. *In:* R.E. Good, D.F. Whigham, and R.L. Simpson (editors). Freshwater Wetlands, Ecological Processes and Management Potential. Academic Press, Inc., New York, pp. 321–340.

Sport Fishing Institute. 1986. Gallup poll points to increased fishing participation. SFI Bulletin 372: 7.

Sugihara, T., C. Yearsley, J.B. Durand, and N.P. Psuty. 1979. Comparison of natural and Altered Estuarine Systems: Analysis. Rutgers Univ., Center for Coastal and Environmental Studies, New Brunswick, NJ. Pub. No. NJ/RU-DEP-11-9-79. 247 pp.

Swift, B.L. 1980. Breeding Bird Habitats in Forested Wetlands of West-Central Massachusetts. M.S. Thesis, Univ. of Massachusetts, Amherst. 90 pp.

Thibodeau, F.R. and B.D. Ostro. 1981. An economic analysis of wetland protection. J. Environ. Manage. 12: 19–30.

U.S. Army Corps of Engineers. 1976. Natural Valley Storage: A Partnership with Nature. New England Division, Waltham, MA.

U.S. Army Corps of Engineers. 1979. Passaic River Basin Study. Plan of Study. New York District. 240 pp.

U.S. Department of the Interior and Department of Commerce. 1982. 1980 National Survey of Fishing, Hunting and Wildlife Associated Recreation. Fish and Wildlife Service and Bureau of Census. 156 pp.

U.S. Water Resources Council. 1977. Estimated Flood Damages. Appendix B. Nationwide Analysis Report. Washington, DC.

U.S. Water Resources Council. 1978. The Nation's Water Resources 1975–2000. Vol. 1: Summary. Washington, DC. 86 pp.

Wander, W. 1980. Breeding birds of southern New Jersey cedar swamps. N.J. Audubon VI(4): 51–65.

Wharton, C.H. 1970. The Southern River Swamp—A Multiple Use Environment. School of Business Administration, Georgia State University. 48 pp.

Whigham, D.F. and R.L. Simpson. 1976. The potential use of freshwater tidal marshes in the management of water quality in the Delaware River. *In:* J. Tourbier and R.W. Peirson, Jr. (editors). Biological Control of Water Pollution. Univ. of Pennsylvania Press. pp. 173–186.

Windom, H.L. 1977. Ability of Salt Marshes to Remove Nutrients and Heavy Metals from Dredged Material Disposal Area Effluents. Technical Rept. D-77-37. U.S. Army Engineers, Waterways Expt. Sta., Vicksburg, MS.

Woodhouse, W.W., E.D. Seneca, and S.W. Broome. 1976. Propagation and Use of *Spartina alterniflora* for Shoreline Erosion Abatement. U.S. Army Coastal Engineering Research Center. Tech. Rept. 76-2.

Wolman, W.G. and L.B. Leopold. 1957. River Floodplains. Some Observations on Their Formation. U.S. Geol. Survey Prof. Paper 282-C.

CHAPTER 8.
Wetland Protection

Introduction

A variety of techniques are available to protect our remaining wetlands, including land-use regulations, direct acquisition, conservation easements, tax incentives, public education, and the efforts of private individuals and corporations. These techniques are discussed in numerous sources including Kusler (1983), Burke and others (1989), and Rusmore and others (1982).

Wetland Regulation

Several Federal and state laws or programs regulate certain uses of Rhode Island wetlands. The more significant ones include the Rivers and Harbors Act of 1899 and the Clean Water Act of 1977 at the Federal level and the Coastal Resources Management Program (1977) and Fresh Water Wetlands Act of 1971 at the state level. Key points of these laws are outlined in Table 21. In addition, Executive Order 11990—"Protection of Wetlands"— requires Federal agencies to develop guidelines to minimize destruction and degradation of wetlands and to preserve and enhance wetland values.

The foundations of Federal wetland regulations are Section 10 of the Rivers and Harbors Act and Section 404 of the Clean Water Act. Federal permits for many types of construction in wetlands are required from the U.S. Army Corps of Engineers, but normal agricultural and silvicultural activities are exempt from permit requirements. The Service plays an active role in the permit process by reviewing permit applications and making recommendations based on environmental considerations, under authority of the Fish and Wildlife Coordination Act. Although the Federal laws in combination apply to virtually all of Rhode Island's wetlands, the U.S. Army Corps of Engineers' 1982 regulations for Section 404 of the Clean Water Act reduced its effectiveness for protecting wetlands. In particular, the widespread use of "nationwide permits" and the lack of strong enforcement were major weak points. Under the nationwide permit system, there was no required reporting or monitoring system, consequently there was no record of wetland loss and no effort to promote environmental or other public interest concerns. In Rhode Island, many wetlands lie above designated headwaters or exist in isolated basins and they were not protected under the 1982 regulations. Numerous lawsuits were filed nationwide against the Corps by concerned environmental organizations over the 1982 regulatory changes. Under an out-of-court settlement agreement (National Wildlife Federation vs. Marsh), the Corps issued regulations in November 1986 requiring closer Federal and state review of proposals to fill wetlands. Implementation of these new regulations needs to be monitored to assess their effectiveness of protecting wetlands.

Wetlands are regulated by the State of Rhode Island under two programs: (1) Coastal Resources Management Program and (2) Fresh Water Wetlands Program. The former program is administered by the Coastal Resources Management Council and deals with a wide range of coastal resources of which coastal wetlands are but one part (Olsen and Seavey 1983). The latter program is administered by the Department of Environmental Management (DEM). Both programs require permits for regulated activities in these wetlands.

Besides the Federal and state permit programs, Section 401 of the Federal Clean Water Act gives the state another powerful tool to protect wetlands. Any Federal permit or license which may involve a discharge to waters of the United States requires a Section 401 water quality certification from the state. The state reviews these permits to see if they meet state water quality standards. If they do not, then 401 certification is denied and the Federal permit cannot be issued. Consequently, DEM has the authority to issue, condition, waive or deny water quality certification for Federal permits including Section 404 permits. This program provides the state with another powerful tool to protect wetlands.

Wetland Acquisition

Wetlands may also be protected by direct acquisition or conservation easements. Many wetlands are owned by public agencies or by private environmental organizations, although the majority are privately-owned.

The U.S. Fish and Wildlife Service's National Wildlife Refuge System was established to preserve important migratory bird wetlands at strategic locations across the country. Four National Wildlife Refuges are located in Rhode Island: Trustom Pond (642 acres), Ninigret (408 acres), Sachuest Point (242 acres), Block Island (46

Table 21. Summary of primary Federal and state laws requiring permits for wetland alteration in Rhode Island.

Name of Law/ Regulation	Administering Agency	Types of Wetlands Regulated	Regulated Activities	Exemptions	Comments
Rivers and Harbors Act of 1899 (Section 10)	U.S. Army Corps of Engineers	Tidal wetlands below the mean high water mark; nontidal wetlands below the ordinary high water mark	Structures and/or work in or affecting the navigable U.S., including dredging and filling	None specified	July 22, 1982 Regulations; U.S. Fish and Wildlife Service and state wildlife agency review permit applications for environmental impacts by authority of Fish and Wildlife Coordination Act.
Clean Water Act of 1977 (Section 404; formerly Federal Water Pollution Control Act of 1972)	U.S. Army Corps of Engineers under guidelines developed by the U.S. Environmental Protection Agency	Wetlands contiguous with all waters of the U.S.	Discharge of dredge or fill material	Normal farming, silviculture, and ranching activities (including minor drainage); maintenance of existing structures; construction or maintenance of farm ponds, irrigation ditches or maintenance of irrigation ditches; construction of temporary sedimentation basins; construction or maintenance of farm roads, forest roads or temporary mining roads (within certain specifications)	July 22, 1982 Regulations; U.S. Environmental Protection Agency oversight; U.S. Fish and Wildlife Service and state wildlife agency review proposed work for environmental impacts by authority of Fish and Wildlife Coordination Act. Permits cannot be issued without State certification that proposed discharge meets State water quality standards. Individual permits are required for specific work in many wetlands; regional permits for certain categories of activities in specified geographic areas; nation-wide permits for 25 specific activities and for discharges into wetlands above headwaters or not part of surface tributary system to interstate or navigable waters of U.S. State takeover of permit program is encouraged. New regulations were issued in October 1984.
State of Rhode Island Coastal Resources Mgmt. Program (as amended June 28, 1983)	Coastal Resources Management Council	Coastal wetlands (salt marshes and contiguous freshwater or brackish wetlands)	In Type 1 (conservation) and Type 2 (low-intensity use) waters, all alterations are prohibited except minimal alterations required by construction or repair of an approved structural shoreline protection facility. In type 2 waters, minor disturbances associated with residential docks and walkways meeting certain standards are permitted. In other waters (types 3, 4, 5, and 6—high-intensity boating, multipurpose, commercial/recreational harbors, and industrial waterfronts/commercial navigation channels, respectively), salt marshes not designated for preservation may be altered if (a) the alteration is made to accomodate a designated priority use for that water area, (b) the applicant has examined all reasonable alternatives and the Council has determined that the selected alternative is the most reasonable, and (c) only the minimum alteration necessary to support the priority use is made.	"Special exceptions" are granted for activities of compelling public purpose (special requirements) that minimize environmental impacts and for which no reasonable alternative is available (subject to proper notice, public hearings and necessary conditions)	Type 1 through 6 waters are shown on maps and coastal wetlands designated for preservation in type 3, 4, 5, and 6 waters are also shown on maps. Field determinations of wetland boundaries are required. The Council may require creation of replacement salt marsh of similar size for any alterations.
Fresh Water Wetlands Act (1971, 1979 amendments)	Department of Environmental Management	Fresh water wetlands (e.g., marshes—1 acre or larger in size; swamps—3 acres or more; ponds—more than ¼ acre)	Drain, fill, excavate, dam, dike, or divert water, place trash, garbage, sewage, highway runoff, drainage ditch effluents and other materials and effluents upon, change or otherwise alter the character of any fresh water wetland	None specified	Regulations also pertain to activities on uplands within 50' of wetland. Activities in rivers and on flood plains and river banks are regulated as fresh water wetlands.

acres), and Pettaquamscutt Cove (26 acres). The State of Rhode Island possesses much wetland acreage. Many wildlife management areas include some large wetland complexes, such as Great Swamp (South Kingstown/Kingston), Burlingame (Charlestown), and Blackhut (Burrillville). Wetlands are also located in various state parks and other conservation areas in Rhode Island (e.g., Audubon Society's Norman Bird Sanctuary in Middletown).

Future Actions

In an effort to maintain and enhance remaining wetlands, many opportunities are available to both government and the private sector. Their joint efforts will determine the future course of our Nation's wetlands. Major options have been outlined below:

Government Options

1. Strengthen Federal, State and local wetlands protection.

2. Ensure proper implementation of existing laws and policies through adequate staffing and improved surveillance and enforcement programs.

3. Increase wetland acquisition in vulnerable areas.

4. Remove government subsidies for wetland drainage.

5. Scrutinize cost-benefit analyses and justifications for flood control projects that involve channelization or other alteration of wetlands and watercourses.

6. Provide tax incentives to private landowners to encourage wetland preservation.

7. Increase support for the Water Bank and Conservation Easement Programs.

8. Increase the number of marsh creation projects, especially related to mitigation for unavoidable wetlands losses by government-sponsored water resource projects; this should include restoration of degraded or former wetlands.

9. Enhance existing wetlands through improving water quality and establishing buffer zones.

10. Monitor wetland changes especially with reference to effectiveness of State and Federal wetland protection efforts and periodically update the National Wetlands Inventory in problem areas.

11. Increase public awareness of wetland values and the status of wetlands through various media and environmental education programs.

12. Conduct research to increase our knowledge of wetland values and ecology.

Private Options:

1. Rather than drain or fill wetlands, seek more environmentally compatible, alternative uses of those areas, e.g., timber harvest, waterfowl production, fur harvest, hay and forage, wild rice production, and hunting leases.

2. Donate wetlands to private or public conservation agencies for tax purposes.

3. Maintain wetlands as open space and seek appropriate tax relief.

4. When selling property that includes wetlands, consider incorporating into the property transfer, a deed restriction or a covenant preventing future alteration and destruction of the wetlands and an appropriate buffer zone.

5. Work in concert with government agencies to help educate the public on wetland values, threats, and losses, for example.

6. Construct ponds in upland areas and manage them for wetland and aquatic species.

7. Purchase Federal and State duck stamps which support wetland acquisition.

8. Support in various ways, public and private efforts to protect and enhance wetlands.

Public and private cooperation is needed to secure a promising future for our remaining wetlands. In Rhode Island, as competition for wetlands between development and environmental interests increases, ways have to be found to achieve economic growth, while minimizing adverse environmental impacts. This is vital to preserving wetland values for our future generations and for fish and wildlife species.

References

Burke, D.G., E.J. Meyers, R.W. Tiner, Jr., and H. Groman. 1989. Protecting Nontidal Wetlands. American Planning Association, Chicago, IL. Planning Advisory Rept. 412/413. 76 pp.

Kusler, J.A. 1983. Our National Wetland Heritage. A Protection Guidebook. Environmental Law Institute, Washington, D.C. 167 pp.

Olsen, S. and G.L. Seavey. 1983. The State of Rhode Island Coastal Resource Management Program as amended on June 28, 1983. Coastal Resources Center, University of Rhode Island, Kingston. 127 pp.

Rusmore, B., A. Swaney, and A.D. Spader (editors). 1982. Private Options: Tools and Concepts for Land Conservation. Island Press, Covelo, CA. 292 pp.

Appendix. List of Plant Species that Occur in Rhode Island's Wetlands

Plant Species that Occur in Rhode Island's Wetlands

GENUS-SPECIES-AUTHOR-TRINOMIAL-AUTHOR	R1IND
ABIES BALSAMEA (L.) MILL	FAC
ACALYPHA RHOMBOIDEA RAF.	FACU-
ACER NEGUNDO L.	FACU-
ACER PENSYLVANICUM L.	FACU
ACER RUBRUM L.	FAC
ACER RUBRUM L. VAR. TRILOBUM TORR. & GRAY EX K. KOCH	FACW+
ACER SACCHARINUM L.	FACW
ACER SACCHARUM MARSHALL	FACU-
ACER SPICATUM LAM.	FACU
ACHILLEA MILLEFOLIUM L.	FACU
ACORUS CALAMUS L.	OBL
ADIANTUM PEDATUM L.	FAC-
AEGOPODIUM PODAGRARIA L.	FACU
AGALINIS MARITIMA (RAF.) RAF.	FACW+
AGALINIS OBTUSIFOLIA (RAF.) PENNELL	FACW+
AGALINIS PAUPERCULA (GRAY) BRITTON	FACW+
AGALINIS PURPUREA (L.) RAF.	FACW-
AGALINIS TENUIFOLIA (VAHL) RAF.	FAC
AGERATINA ALTISSIMA (L.) R.M. KING & H. ROB.	FACU-
AGRIMONIA GRYPOSEPALA WALLR.	FACU
AGRIMONIA STRIATA MICHX.	FACU
AGROPYRON CANINUM BEAUV.	FACU-
AGROPYRON REPENS (L.) BEAUV.	FACU
AGROPYRON TRACHYCAULUM (LINK) MALTE EX H.F. LEWIS	FACW
AGROSTIS ALBA L.	FACW
AGROSTIS CANINA L.	FACU
AGROSTIS GIGANTEA ROTH	NI
AGROSTIS HYEMALIS (WALTER) B.S.P.	FAC
AGROSTIS PERENNANS (WALTER) TUCKERMAN	FACU
AGROSTIS SCABRA WILLD.	FAC
AGROSTIS STOLONIFERA L.	FACW
AILANTHUS ALTISSIMA (MILL.) SWINGLE	NI
ALETRIS FARINOSA L.	FAC
ALISMA PLANTAGO-AQUATICA L.	OBL
ALISMA SUBCORDATUM RAF.	OBL
ALLIUM CANADENSE L.	FACU
ALLIUM TRICOCCUM AIT.	FACU+
ALLIUM VINEALE L.	FACU-
ALNUS GLUTINOSA (L.) GAERTN.	FACW-
ALNUS INCANA (L.) MOENCH	NI
ALNUS MARITIMA (MARSH.) MUHL	OBL
ALNUS RUGOSA (DU ROI) SPRENG.	FACW+
ALNUS SERRULATA (AIT.) WILLD.	OBL
ALOPECURUS GENICULATUS L.	OBL
ALOPECURUS MYOSUROIDES HUDS.	FACW
ALOPECURUS PRATENSIS L.	FACW
ALTHAEA OFFICINALIS L.	FACW+
AMARANTHUS ALBUS L.	FACU
AMARANTHUS BLITOIDES S. WATS.	NI
AMARANTHUS CANNABINUS (L.) SAUER	OBL
AMARANTHUS PUMILUS RAF.	FACW*

GENUS-SPECIES-AUTHOR-TRINOMIAL AUTHOR	R1IND
AMARANTHUS RETROFLEXUS L.	FACU
AMARANTHUS SPINOSUS L.	FACU
AMBROSIA ARTEMISIIFOLIA L.	FACU-
AMBROSIA TRIFIDA L.	FAC-
AMELANCHIER ARBOREA (MICHX. F.) FERN.	FAC
AMELANCHIER CANADENSIS (L.) MEDIC.	FACU-
AMELANCHIER SPICATA (LAM.) K. KOCH	FACU
AMMOPHILA BREVILIGULATA FERNALD	FACU-
AMORPHA FRUTICOSA L.	FACW
AMPHICARPAEA BRACTEATA (L.) FERNALD	FAC
ANDROMEDA GLAUCOPHYLLA LINK	OBL
ANDROMEDA POLIFOLIA L.	OBL
ANDROPOGON GERARDI VITMAN	FAC
ANDROPOGON GLOMERATUS (WALTER) B.S.P.	FACW+
ANDROPOGON VIRGINICUS L.	FACU
ANEMONE QUINQUEFOLIA L.	FACU
ANEMONE RIPARIA FERNALD	NI
ANEMONE VIRGINIANA L.	NI
ANGELICA ATROPURPUREA L.	OBL
ANGELICA LUCIDA L.	FAC*
ANTHEMIS COTULA L.	FACU-
ANTHOXANTHUM ODORATUM L.	FACU
APIOS AMERICANA MEDIC.	FACW
APOCYNUM CANNABINUM L.	FACU
AQUILEGIA CANADENSIS L.	FACU
ARABIS DRUMMONDII GRAY	FAC
ARALIA NUDICAULIS L.	FACU
ARALIA SPINOSA L.	FAC
ARCTOSTAPHYLOS UVA-URSI (L.) SPRENG.	FACU
ARENARIA SERPYLLIFOLIA L.	NI
ARETHUSA BULBOSA L.	FAC
ARISAEMA DRACONTIUM (L.) SCHOTT	OBL
ARISAEMA TRIPHYLLUM (L.) SCHOTT	FACW
ARMORACIA RUSTICANA P. GAERTN. B. MEYER & SCHERB.	FACW-
ARONIA ARBUTIFOLIA (L.) ELLIOTT	NI
ARONIA MELANOCARPA (MICHX.) ELLIOTT	FACW
ARONIA PRUNIFOLIA (MARSH.) REHDER	FAC
ARRHENATHERUM ELATIUS (L.) J. & K. PRESL	FACU
ARTEMISIA BIENNIS WILLD.	FAC
ARTEMISIA STELLERANA BESSER	FACU-
ASCLEPIAS EXALTATA L.	FACU*
ASCLEPIAS INCARNATA L.	OBL
ASCLEPIAS PURPURASCENS L.	FACU
ASPARAGUS OFFICINALIS L.	FACU
ASPLENIUM PLATYNEURON (L.) OAKES	FACU
ASTER DUMOSUS L.	FAC
ASTER ERICOIDES L.	FACU
ASTER JUNCIFORMIS RYDB.	OBL
ASTER LATERIFLORUS (L.) BRITTON	FACW-
ASTER LUCIDULUS (GRAY) WIEGAND	FACW
ASTER NEMORALIS AIT.	FACW+

Symbology: OBL (Obligate), FACW (Facultative Wetland), FAC (Facultative), FACU (Facultative Upland), NI (no indicator assigned), * (limited ecological information), + (higher portion of frequency range), and − (lower portion of frequency range). See discussion of hydrophyte definition and concept in Chapter 6.

A-1

Plant Species that Occur in Rhode Island's Wetlands (Continued)

GENUS-SPECIES-AUTHOR-TRINOMIAL-TRINOMIAL AUTHOR	R1IND	GENUS-SPECIES-AUTHOR-TRINOMIAL-TRINOMIAL AUTHOR	R1IND
ASTER NOVAE-ANGLIAE L.	FACW-	CABOMBA CAROLINIANA GRAY	OBL
ASTER NOVI-BELGII L.	FACW+	CAKILE EDENTULA (BIGEL.) HOOK.	FACU
ASTER PRAEALTUS POIR.	FACW	CALAMAGROSTIS CANADENSIS (MICHX.) BEAUV.	FACW+
ASTER PUNICEUS L.	OBL	CALAMAGROSTIS CINNOIDES (MUHL) BARTON	OBL
ASTER RADULA AIT.	FACW	CALLA PALUSTRIS L.	OBL
ASTER SIMPLEX WILLD.	OBL	CALLITRICHE HETEROPHYLLA PURSH	OBL
ASTER SUBULATUS MICHX.	OBL	CALLITRICHE VERNA L.	FAC*
ASTER TENUIFOLIUS L.	FACW	CALLUNA VULGARIS (L.) HULL	FACW+
ASTER TRADESCANTI L.	FACW	CALOPOGON TUBEROSUS (L.) B.S.P.	OBL
ASTER UMBELLATUS MILL.	FACW	CALTHA PALUSTRIS L.	OBL
ASTER VIMINEUS LAM.	FAC	CALYSTEGIA SEPIUM (L.) R. BR.	FAC-
ASTER X BLAKEI (T. PORTER) HOUSE	FACW+	CAMPANULA APARINOIDES PURSH	OBL
ASTER X LANCEOLATUS WILLD.	NI	CANNABIS SATIVA L.	FACU
ATHYRIUM FILIX-FEMINA (L.) ROTH	FAC	CAPSELLA BURSA-PASTORIS (L.) MEDIC.	FACU
ATHYRIUM THELYPTEROIDES (MICHX.) DESV.	FAC	CARDAMINE BULBOSA (SCHREB. EX MUHL) B.S.P.	OBL
ATRIPLEX ARENARIA NUTT.	FAC-	CARDAMINE HIRSUTA L.	FACU
ATRIPLEX GLABRIUSCULA EDMONST.	NI	CARDAMINE PARVIFLORA L.	FACU
ATRIPLEX PATULA L.	FACW	CARDAMINE PENSYLVANICA MUHL. EX WILLD.	OBL
BACCHARIS HALIMIFOLIA L.	FACW	CARDAMINE PRATENSIS L.	OBL
BARBAREA VULGARIS R. BR.	FACU	CAREX ABSCONDITA MACKENZ.	FAC
BARTONIA PANICULATA (MICHX.) MUHL	OBL	CAREX ALBOLUTESCENS SCHWEINITZ	FACW
BARTONIA VIRGINICA (L.) B.S.P.	FACW	CAREX AMPHIBOLA STEUD.	FACW
BERBERIS THUNBERGII DC.	FACU	CAREX ANNECTENS (BICKN.) BICKN.	FAC
BERBERIS VULGARIS L.	FACU	CAREX ATLANTICA L.H. BAILEY	FACW+
BETULA ALLEGHANIENSIS BRITTON	FAC	CAREX AUREA NUTT.	FACW
BETULA LENTA L.	FACU	CAREX BICKNELLII BRITTON	FACU
BETULA NIGRA L.	FACW	CAREX BLANDA DEWEY	FAC
BETULA PAPYRIFERA MARSHALL	FACU	CAREX BROMOIDES SCHKUHR	FACW
BETULA POPULIFOLIA MARSHALL	FAC	CAREX BRUNNESCENS (PERS.) POIR.	FACW
BIDENS CERNUA L.	OBL	CAREX BULLATA SCHKUHR	OBL
BIDENS COMOSA (GRAY) WIEGAND	FACW	CAREX BUSHII MACKENZ.	FACW
BIDENS CONNATA MUHL. EX WILLD.	FACW	CAREX BUXBAUMII WAHLENB.	OBL
BIDENS CORONATA (L.) BRITTON	FACW+	CAREX CANESCENS L.	OBL
BIDENS DISCOIDEA (TORR. & GRAY) BRITTON	OBL	CAREX CEPHALOIDEA DEWEY	FAC+
BIDENS EATONI FERNALD	FACW	CAREX CEPHALOPHORA MUHL. EX WILLD.	FACU
BIDENS FRONDOSA L.	FACW	CAREX COLLINSII NUTT.	OBL
BIDENS LAEVIS (L.) B.S.P.	OBL	CAREX COMOSA BOOTT	OBL
BIDENS TRIPARTITA L.	OBL	CAREX CONOIDEA SCHKUHR	FACU
BOEHMERIA CYLINDRICA (L.) SWARTZ	FACW+	CAREX CRANEI DEWEY	FACW
BOLTONIA ASTEROIDES (L.) L'HER.	FACW	CAREX CRAWFORDII FERNALD	FAC
BOTRYCHIUM DISSECTUM SPRENG.	FAC	CAREX CRINITA LAM.	OBL
BOTRYCHIUM LANCEOLATUM (S.G. GMEL.) RUPR.	FACW	CAREX CRISTATELLA BRITTON	FACW
BOTRYCHIUM LUNARIA (L.) SWARTZ	FACW	CAREX CRYPTOLEPIS MACKENZ.	OBL
BOTRYCHIUM MATRICARIIFOLIUM A. BRAUN	FACU	CAREX DEBILIS MICHX.	FAC
BOTRYCHIUM SIMPLEX E. HITCHC.	FACU	CAREX DEWEYANA SCHWEINITZ	FACU
BRASENIA SCHREBERI J.F. GMEL.	OBL	CAREX DIANDRA SCHRANK	OBL
BRIZA MEDIA L.	FAC	CAREX ECHINATA MURRAY	OBL*
BROMUS CILIATUS L.	FAC+	CAREX EXILIS DEWEY	OBL
BROMUS DUDLEYI FERNALD	FACW	CAREX FLAVA L.	NI
BROMUS LATIGLUMIS (SHEAR) HITCHC.	FACW	CAREX FOENEA WILLD.	FACU*
BULBOSTYLIS CAPILLARIS (L.) C.B. CLARKE	FACU	CAREX GRACILLIMA SCHWEINITZ	OBL
		CAREX HAYDENII DEWEY	

Symbology: OBL (Obligate), FACW (Facultative Wetland), FAC (Facultative), FACU (Facultative Upland), NI (no indicator assigned), * (limited ecological information), + (higher portion of frequency range), and − (lower portion of frequency range). See discussion of hydrophyte definition and concept in Chapter 6.

Plant Species that Occur in Rhode Island's Wetlands (Continued)

GENUS-SPECIES-AUTHOR-TRINOMIAL-TRINOMIAL AUTHOR	RIIND	GENUS-SPECIES-AUTHOR-TRINOMIAL-TRINOMIAL AUTHOR	RIIND
CAREX HORMATHODES FERNALD	OBL	CASSIA HEBECARPA FERNALD	FAC
CAREX HOWEI MACKENZ.	OBL	CASSIA NICTITANS L.	FACU-
CAREX HYSTERICINA MUHL EX WILLD.	OBL	CASTILLEJA COCCINEA (L.) SPRENG.	FAC
CAREX INTERIOR L.H. BAILEY	OBL	CELASTRUS SCANDENS L.	FACU-
CAREX INTUMESCENS RUDGE	FACW+	CELTIS OCCIDENTALIS L.	FACU
CAREX LACUSTRIS WILLD.	OBL	CENTAURIUM UMBELLATUM GILIB. EX FERNALD	FAC-
CAREX LAEVIVAGINATA (KUEKENTH.) MACKENZ.	OBL	CEPHALANTHUS OCCIDENTALIS L.	OBL
CAREX LASIOCARPA EHRH.	OBL	CERASTIUM VULGATUM L.	FACU-
CAREX LAXIFLORA LAM.	FACU*	CERATOPHYLLUM DEMERSUM L.	OBL
CAREX LEPIDOCARPA TAUSCH	OBL	CERATOPHYLLUM MURICATUM CHAM.	OBL
CAREX LEPTALEA WAHLENB.	OBL	CHAMAECYPARIS THYOIDES (L.) B.S.P.	OBL
CAREX LIMOSA L.	OBL	CHAMAEDAPHNE CALYCULATA (L.) MOENCH	OBL
CAREX LONGII MACKENZ.	OBL	CHELONE GLABRA L.	OBL
CAREX LUPULINA MUHL EX WILLD.	OBL	CHENOPODIUM ALBUM L.	FACU+
CAREX LURIDA WAHLENB.	OBL	CHENOPODIUM AMBROSIOIDES L.	FACU
CAREX MEADII DEWEY	FAC	CHENOPODIUM GLAUCUM L.	FACW-
CAREX NIGRA (L.) REICHARD	FACW+	CHENOPODIUM LEPTOPHYLLUM (MOQ.) NUTT. EX S. WATS.	FAC
CAREX NORMALIS MACKENZ.	FACU	CHENOPODIUM RUBRUM L.	FACW
CAREX NOVAE-ANGLIAE SCHWEINITZ	FACU*	CHRYSOSPLENIUM AMERICANUM SCHWEINITZ	OBL
CAREX PAUPERCULA MICHX.	OBL	CICUTA BULBIFERA L.	OBL
CAREX POLYMORPHA MUHL	FACU	CICUTA MACULATA L.	OBL
CAREX PRASINA WAHLENB.	OBL	CINNA ARUNDINACEA L.	FACW+
CAREX PSEUDOCYPERUS L.	OBL	CIRCAEA ALPINA L.	FACW
CAREX RETRORSA SCHWEINITZ	FACW+	CIRCAEA LUTETIANA L.	FACU
CAREX ROSTRATA J. STOKES	OBL	CIRSIUM ARVENSE (L.) SCOP.	FACU
CAREX SCABRATA SCHWEINITZ	OBL	CIRSIUM HORRIDULUM MICHX.	FACU-
CAREX SCHWEINITZII DEWEY	OBL	CIRSIUM MUTICUM MICHX.	OBL
CAREX SCOPARIA SCHKUHR EX WILLD.	FACW	CIRSIUM VULGARE (SAVI) TENORE	FACU-
CAREX SEORSA E.C. HOWE	FACW	CLADIUM MARISCOIDES (MUHL) TORR.	OBL
CAREX SPARGANIOIDES MUHL EX WILLD.	FACU	CLAYTONIA VIRGINICA L.	FACU
CAREX SQUARROSA L.	FACW	CLEMATIS VIRGINIANA L.	FAC+
CAREX STRAMINEA WILLD.	OBL	CLETHRA ALNIFOLIA L.	FAC
CAREX STRICTA LAM.	OBL	CLINTONIA BOREALIS (AIT.) RAF.	FAC+
CAREX SWANII (FERNALD) MACKENZ.	FACU	COELOGLOSSUM VIRIDE (L.) HARTM.	FACU
CAREX TENERA DEWEY	FAC	COLLINSONIA CANADENSIS L.	FAC+
CAREX TORTA BOOTT	FACW	COMANDRA UMBELLATA (L.) NUTT.	FACU-
CAREX TRIBULOIDES WAHLENB.	FACW+	COMMELINA COMMUNIS L.	FAC-
CAREX TRISPERMA DEWEY	OBL	CONIUM MACULATUM L.	FACW
CAREX TUCKERMANII BOOTT	OBL	COPTIS TRIFOLIA (L.) SALISB.	FACW
CAREX VESICARIA L.	OBL	CORALLORHIZA MACULATA (RAF.) RAF.	FACU
CAREX VULPINOIDEA MICHX.	OBL	CORALLORHIZA TRIFIDA CHAT.	FACW
CAREX WALTERANA L.H. BAILEY	OBL	COREOPSIS LANCEOLATA L.	FACW
CAREX WIEGANDII MACKENZ.	OBL	COREOPSIS ROSEA NUTT.	FAC-
CAREX X ALATA TORR.	OBL	COREOPSIS TINCTORIA NUTT.	FAC-
CAREX X STIPATA MUHL EX WILLD.	FAC	CORISPERMUM HYSSOPIFOLIUM L.	FACW
CARPINUS CAROLINIANA WALTER	FACU+	CORNUS AMOMUM MILL	FACW
CARYA CORDIFORMIS (WANGENH.) K. KOCH	FACU-	CORNUS CANADENSIS L.	FAC-
CARYA GLABRA (MILL.) SWEET	FACU-	CORNUS FLORIDA L.	FACU-
CARYA LACINIOSA (MICHX. F.) LOUD.	FAC	CORNUS FOEMINA MILL.	FAC
CARYA OVALIS (WANGENH.) SARG.	NI	CORNUS STOLONIFERA MICHX.	FACW+
CARYA OVATA (MILL.) K. KOCH	FACU-	CORYLUS AMERICANA WALTER	FACU-
CASSIA FASCICULATA MICHX.	FACU	CORYLUS CORNUTA MARSHALL	FACU-

Symbology: OBL (Obligate), FACW (Facultative Wetland), FAC (Facultative), FACU (Facultative Upland), NI (no indicator assigned), * (limited ecological information), + (higher portion of frequency range), and — (lower portion of frequency range). See discussion of hydrophyte definition and concept in Chapter 6.

Plant Species that Occur in Rhode Island's Wetlands (Continued)

GENUS-SPECIES-AUTHOR-TRINOMIAL-TRINOMIAL AUTHOR	R1IND	GENUS-SPECIES-AUTHOR-TRINOMIAL-TRINOMIAL AUTHOR	R1IND
CRATAEGUS CRUS-GALLI L.	FACU	DRYOPTERIS X TRIPLOIDEA WHERRY	FAC
CRATAEGUS PHAENOPYRUM (L.F.) MEDIC.	FAC	DULICHIUM ARUNDINACEUM (L.) BRITTON	OBL
CRYPTOTAENIA CANADENSIS (L.) DC.	FAC-	ECHINOCHLOA CRUSGALLI (L.) BEAUV.	FACU
CUPHEA VISCOSISSIMA JACQ.	FAC-	ECHINOCHLOA MURICATA (BEAUV.) FERNALD	FACW+
CYPERUS ARISTATUS ROTTB.	FACW+	ECHINOCHLOA WALTERI (PURSH) A. HELLER	FACW+
CYPERUS DENTATUS TORR.	FACW	ELATINE AMERICANA (PURSH) ARN.	FAC
CYPERUS DIANDRUS TORR.	FACW+	ELATINE MINIMA (NUTT.) FISCH. & C.A. MEYER	OBL
CYPERUS ERYTHRORHIZOS MUHL.	FACW	ELEOCHARIS ACICULARIS (L.) ROEM. & J.A. SCHULTES	OBL
CYPERUS ESCULENTUS L.	FACW	ELEOCHARIS ENGELMANNI STEUD.	OBL
CYPERUS FILICINUS VAHL	OBL	ELEOCHARIS EQUISETOIDES (ELLIOTT) TORR.	FACW+
CYPERUS FLAVESCENS L.	OBL	ELEOCHARIS ERYTHROPODA STEUD.	OBL
CYPERUS ODORATUS L.	FACW+	ELEOCHARIS FALLAX WEATHERBY	OBL
CYPERUS RIVULARIS KUNTH	FACW	ELEOCHARIS HALOPHILA (FERNALD & A. BRACKETT) FERNALD & A. BRACKETT	OBL
CYPERUS STRIGOSUS L.	FACU	ELEOCHARIS MELANOCARPA TORR.	FACW+
CYPRIPEDIUM ACAULE AIT.	FAC+	ELEOCHARIS OBTUSA (WILLD.) J.A. SCHULTES	OBL
CYPRIPEDIUM CALCEOLUS L.	FAC	ELEOCHARIS OLIVACEA TORR.	OBL
CYSTOPTERIS BULBIFERA (L.) BERNH.	FACU	ELEOCHARIS OVATA (ROTH) ROEM. & J.A. SCHULTES	OBL
CYSTOPTERIS FRAGILIS (L.) BERNH.	FACU	ELEOCHARIS PALUSTRIS (L.) ROEM. & J.A. SCHULTES	OBL
DACTYLIS GLOMERATA L.	FACU	ELEOCHARIS PARVULA (ROEM. & J.A. SCHULTES) LINK EX BLUFF & FINGERH.	OBL
DALIBARDA REPENS L.	FAC	ELEOCHARIS PAUCIFLORA (LIGHTF.) LINK	OBL
DANTHONIA COMPRESSA AUST.	FACU-	ELEOCHARIS ROBBINSII OAKES	OBL
DANTHONIA SERICEA NUTT.	FACU	ELEOCHARIS ROSTELLATA (TORR.) TORR.	OBL
DECODON VERTICILLATUS (L.) ELLIOTT	OBL	ELEOCHARIS SMALLII BRITTON	OBL
DESCHAMPSIA CESPITOSA (L.) BEAUV.	FACW	ELEOCHARIS TENUIS (WILLD.) J.A. SCHULTES	FACW+
DESMODIUM CANADENSE (L.) DC.	FAC	ELEOCHARIS TRICOSTATA TORR.	OBL
DICHANTHELIUM ACUMINATUM (SWARTZ) GOULD & C.A. CLARK	FAC+	ELEOCHARIS TUBERCULOSA (MICHX.) ROEM. & J.A. SCHULTES	OBL
DICHANTHELIUM CLANDESTINUM (J.A. SCHULTES) GOULD	FACU+	ELEOCHARIS UNIGLUMIS (LINK) J.A. SCHULTES	OBL
DICHANTHELIUM COMMUTATUM (L.) GOULD	FAC	ELEUSINE INDICA (L.) GAERTN.	FACU-
DICHANTHELIUM DICHOTOMUM (L.) GOULD	FACU	ELODEA CANADENSIS MICHX.	OBL
DICHANTHELIUM LATIFOLIUM (L.) GOULD	FACU	ELODEA NUTTALLII (PLANCH.) H. ST. JOHN	OBL
DICHANTHELIUM LAXIFLORUM (L.) GOULD	FACU	ELYMUS CANADENSIS L.	FACU+
DICHANTHELIUM OLIGOSANTHES (J.A. SCHULTES) GOULD	FACU	ELYMUS RIPARIUS WIEGAND	FACW
DICHANTHELIUM SABULORUM (LAM.) GOULD & C.A. CLARK	OBL	ELYMUS VILLOSUS MUHL. EX WILLD.	FACU-
DICHANTHELIUM SCABRIUSCULUM (ELLIOTT) GOULD & C.A. CLARK	FACW	ELYMUS VIRGINICUS L.	FACW-
DICHANTHELIUM SCOPARIUM (LAM.) GOULD	FACU	EPILOBIUM ANGUSTIFOLIUM L.	FAC
DICHANTHELIUM SPHAEROCARPON (ELLIOTT) GOULD	FACU-	EPILOBIUM COLORATUM BIEHLER	OBL
DIGITARIA SANGUINALIS (L.) SCOP.	FAC+	EPILOBIUM HIRSUTUM L.	FACW
DIOSCOREA VILLOSA L.	FAC-	EPILOBIUM LEPTOPHYLLUM RAF.	OBL
DIOSPYROS VIRGINIANA L.	N1	EPILOBIUM PALUSTRE L.	OBL
DIPSACUS SYLVESTRIS HUDS.	FAC	EPILOBIUM STRICTUM MUHL. EX SPRENG.	FACW
DIRCA PALUSTRIS L.	FACW+	EQUISETUM ARVENSE L.	FACW
DISTICHLIS SPICATA (L.) GREENE	FACU-	EQUISETUM FLUVIATILE L.	OBL
DRACOCEPHALUM PARVIFLORUM NUTT.	OBL	EQUISETUM HYEMALE L.	FAC
DROSERA FILIFORMIS RAF.	OBL	EQUISETUM PALUSTRE L.	OBL
DROSERA INTERMEDIA HAYNE	OBL	EQUISETUM SYLVATICUM L.	FACW
DROSERA ROTUNDIFOLIA L.	FACW+	ERAGROSTIS CILIANENSIS (ALL.) LINK EX MOSHER	FACU
DRYOPTERIS CLINTONIANA (D.C. EAT.) P. DOWEL	FACW+	ERAGROSTIS PECTINACEA (MICHX.) NEES	FACU
DRYOPTERIS CRISTATA (L.) GRAY	FACU	ERECHTITES HIERACIIFOLIA (L.) RAF. EX DC.	FAC
DRYOPTERIS INTERMEDIA (WILLD.) GRAY	FACU-	ERIGERON ANNUUS (L.) PERS.	FACU
DRYOPTERIS MARGINALIS (L.) GRAY	FAC+	ERIGERON PHILADELPHICUS L.	FACU
DRYOPTERIS SPINULOSA (O.F. MUELL.) WATT	FACW	ERIGERON PULCHELLUS MICHX.	FACU
DRYOPTERIS X BOOTTII (TUCKERMAN) UNDERW.			

Symbology: OBL (Obligate), FACW (Facultative Wetland), FAC (Facultative), FACU (Facultative Upland), N1 (no indicator assigned), * (limited ecological information), + (higher portion of frequency range), and − (lower portion of frequency range). See discussion of hydrophyte definition and concept in Chapter 6.

Plant Species that Occur in Rhode Island's Wetlands (Continued)

GENUS-SPECIES-AUTHOR-TRINOMIAL-AUTHOR	R1IND
ERIGERON STRIGOSUS MUHL. EX WILLD.	FACU+
ERIOCAULON PARKERI B. ROB.	OBL
ERIOCAULON SEPTANGULARE WITH.	OBL
ERIOPHORUM ALPINUM L.	OBL
ERIOPHORUM ANGUSTIFOLIUM HONCK.	OBL
ERIOPHORUM GRACILE W. KOCH	OBL
ERIOPHORUM SPISSUM FERNALD	OBL
ERIOPHORUM TENELLUM NUTT.	OBL
ERIOPHORUM VIRGINICUM L.	OBL
ERIOPHORUM VAGINATUM L.	OBL
ERIOPHORUM VIRIDICARINATUM (ENGELM.) FERNALD	OBL
ERYSIMUM CHEIRANTHOIDES L.	FACU
EUONYMUS ATROPURPUREUS JACQ.	FACU
EUPATORIADELPHUS DUBIUS (WILLD. EX POIR.) R.M. KING & H. ROB.	FACW
EUPATORIADELPHUS FISTULOSUS (BARRATT EX HOOK.) R.M. KING & H. ROB.	FACW
EUPATORIADELPHUS MACULATUS (L.) R.M. KING & H. ROB.	FACW
EUPATORIUM LEUCOLEPIS (DC.) TORR. & GRAY	FACW+
EUPATORIUM PERFOLIATUM L.	FACW+
EUPATORIUM PILOSUM WALTER	FACW
EUPATORIUM ROTUNDIFOLIUM L.	FAC-
EUPATORIUM SEROTINUM MICHX.	FAC-
EUPHORBIA MACULATA L.	FACU-
EUTHAMIA POLYGONIFOLIA L.	FACU
EUTHAMIA GRAMINIFOLIA (L.) NUTT.	FAC
FAGUS GRANDIFOLIA EHRH.	FACU5
FESTUCA ARUNDINACEA SCHREB.	FACU
FESTUCA OBTUSA BIEHLER	FACU
FESTUCA PRATENSIS HUDS.	FACU-
FESTUCA RUBRA L.	FACU
FIMBRISTYLIS AUTUMNALIS (L.) ROEM. & J.A. SCHULTES	FACW+
FRAGARIA VIRGINIANA DUCHESNE	FACU
FRAXINUS AMERICANA L.	FACU
FRAXINUS NIGRA MARSHALL	FACW
FRAXINUS PENNSYLVANICA MARSHALL	FACW
FUIRENA PUMILA TORR.	OBL
GALIUM APARINE L.	FACU
GALIUM ASPRELLUM MICHX.	OBL
GALIUM OBTUSUM BIGEL.	FACW+
GALIUM PALUSTRE L.	OBL
GALIUM TINCTORIUM L.	OBL
GALIUM TRIFIDUM L.	FACW+
GALIUM TRIFLORUM MICHX.	FACU
GAULTHERIA HISPIDULA (L.) MUHL. EX TORR.	FACW
GAULTHERIA PROCUMBENS L.	FACU
GAYLUSSACIA BACCATA (WANGENH.) K. KOCH	FACU
GAYLUSSACIA DUMOSA (ANDR.) TORR. & GRAY	FAC
GAYLUSSACIA FRONDOSA (L.) TORR. & GRAY	FAC
GENTIANA ANDREWSII GRISEB.	FACW
GENTIANA CLAUSA RAF.	FACW
GENTIANA SAPONARIA L.	FACW
GENTIANOPSIS CRINITA (FROEL.) MA	OBL

GENUS-SPECIES-AUTHOR-TRINOMIAL-AUTHOR	R1IND
GERANIUM MACULATUM L.	FACU
GEUM ALEPPICUM JACQ.	FAC
GEUM CANADENSE JACQ.	FACU
GEUM LACINIATUM MURRAY	FAC+
GEUM RIVALE L.	OBL
GEUM VIRGINIANUM L.	FAC-
GLAUX MARITIMA L.	OBL
GLECOMA HEDERACEA L.	FACU
GLEDITSIA TRIACANTHOS L.	FAC-
GLYCERIA ACUTIFLORA TORR.	OBL
GLYCERIA BOREALIS (NASH) BATCH.	OBL
GLYCERIA CANADENSIS (MICHX.) TRIN.	OBL
GLYCERIA FLUITANS (L.) R. BR.	OBL
GLYCERIA MAXIMA (HARTM.) O.R. HOLMBERG	OBL
GLYCERIA OBTUSA (MUHL.) TRIN.	OBL
GLYCERIA SEPTENTRIONALIS A. HITCHC.	OBL
GLYCERIA STRIATA (LAM.) A. HITCHC.	OBL
GOODYERA PUBESCENS (WILLD.) R. BR.	FACU-
GOODYERA REPENS (L.) R. BR. IN W.T. AIT.	FACU+
GOODYERA TESSELATA LODDIG.	FACU-
GRATIOLA AUREA PURSH	OBL
GRATIOLA NEGLECTA TORR.	OBL
GRINDELIA SQUARROSA (PURSH) DUNAL	FACU
HACKELIA VIRGINIANA (L.) I. JOHNST.	FACU
HAMAMELIS VIRGINIANA L.	FAC-
HASTEOLA SUAVEOLENS (L.) POJARK.	FAC-
HELENIUM AMARUM (RAF.) H. ROCK	FACW-
HELENIUM AUTUMNALE L.	FACW+
HELENIUM FLEXUOSUM RAF.	FAC-
HELIANTHUS ANNUUS L.	FAC-
HELIANTHUS DECAPETALUS L.	FACU
HELIANTHUS GIGANTEUS L.	FACW
HELIANTHUS TUBEROSUS L.	FAC
HEMICARPHA MICRANTHA (VAHL) PAX	FACW+
HERACLEUM LANATUM MICHX.	FACU-
HIBISCUS MOSCHEUTOS L.	OBL
HIEROCHLOE ODORATA (L.) BEAUV.	FACU
HOLCUS LANATUS L.	FACU
HONKENYA PEPLOIDES (L.) EHRH.	FACU
HORDEUM JUBATUM L.	FAC
HOTTONIA INFLATA ELLIOTT	OBL
HOUSTONIA CAERULEA L.	FACU
HUMULUS JAPONICUS SIEBOLD & ZUCCAR.	FACU
HUMULUS LUPULUS L.	NI
HYDROCOTYLE AMERICANA L.	OBL
HYDROCOTYLE UMBELLATA L.	OBL
HYDROCOTYLE VERTICILLATA THUNB.	OBL
HYPERICUM ADPRESSUM W. BARTON	OBL
HYPERICUM BOREALE (BRITTON) BICKN.	OBL
HYPERICUM CANADENSE L.	FACW
HYPERICUM DISSIMULATUM BICKN.	FACW
HYPERICUM ELLIPTICUM HOOK.	OBL

Symbology: OBL (Obligate), FACW (Facultative Wetland), FAC (Facultative), FACU (Facultative Upland), NI (no indicator assigned). * (limited ecological information), + (higher portion of frequency range), and − (lower portion of frequency range). See discussion of hydrophyte definition and concept in Chapter 6.

Plant Species that Occur in Rhode Island's Wetlands (Continued)

GENUS-SPECIES-AUTHOR-TRINOMIAL-TRINOMIAL AUTHOR	RI IND	GENUS-SPECIES-AUTHOR-TRINOMIAL-TRINOMIAL AUTHOR	RI IND
HYPERICUM MAJUS (GRAY) BRITTON	FACW	LACTUCA SERRIOLA L.	FAC-
HYPERICUM MUTILUM L.	FACW	LAPORTEA CANADENSIS (L.) WEDD.	FACW
HYPERICUM PROLIFICUM L.	FACU	LARIX LARICINA (DU ROI) K. KOCH	FACW
HYPERICUM PUNCTATUM LAM.	FAC-	LATHYRUS JAPONICUS WILLD.	FACU-
HYPOXIS HIRSUTA (L.) COVILLE	FAC	LATHYRUS PALUSTRIS L.	FACW+
ILEX GLABRA (L.) GRAY	FACW-	LEERSIA ORYZOIDES (L.) SWARTZ	OBL
ILEX LAEVIGATA (PURSH) A. GRAY	OBL	LEERSIA VIRGINICA WILLD.	FACW
ILEX OPACA SOLAND. IN AIT.	FACW+	LEMNA MINIMA HUMB. EX PHILIPPI NOMEN SUPERFL NONCHEV.	OBL
ILEX VERTICILLATA (L.) GRAY	FACW+	LEMNA MINOR L.	OBL
IMPATIENS CAPENSIS MEERB.	FACW	LEMNA PERPUSILLA TORR.	OBL
IMPATIENS PALLIDA NUTT.	FACW	LEMNA TRISULCA L.	OBL
IPOMOEA COCCINEA L.	FACU	LEMNA VALDIVIANA PHILIPPI	OBL
IRIS PRISMATICA PURSH	OBL	LEPIDIUM DENSIFLORUM SCHRAD.	FAC
IRIS PSEUDACORUS L.	OBL	LEPIDIUM VIRGINICUM L.	FACW-
IRIS VERSICOLOR L.	OBL	LEPTOCHLOA FASCICULARIS (LAM.) GRAY	FACW
ISOETES ECHINOSPORA DURIEU	OBL	LESPEDEZA ANGUSTIFOLIA (PURSH) ELLIOTT	FAC
ISOETES ENGELMANNII A. BRAUN	OBL	LEUCOTHOE RACEMOSA (L.) GRAY	FACW-
ISOETES RIPARIA ENGELM. EX A. BRAUN	OBL	LIATRIS SPICATA (L.) WILLD.	FACW
ISOETES TUCKERMANII A. BRAUN EX ENGELM.	OBL	LIGUSTICUM SCOTHICUM L.	FAC+
ISOTRIA MEDEOLOIDES (PURSH) RAF.	FACU	LIGUSTRUM VULGARE L.	FAC
ISOTRIA VERTICILLATA (MUHL. EX WILLD.) RAF.	FACU	LILAEOPSIS CHINENSIS (L.) KUNTZE	FACW
IVA FRUTESCENS L.	FACW+	LILIUM CANADENSE L.	OBL
JUGLANS CINEREA L.	FACU+	LILIUM PHILADELPHICUM L.	FAC+
JUGLANS NIGRA L.	FACU	LILIUM SUPERBUM L.	FACU+
JUNCUS ACUMINATUS MICHX.	OBL	LIMONIUM CAROLINIANUM (WALTER) BRITTON	FACW+
JUNCUS ARTICULATUS L.	OBL	LIMONIUM NASHII SMALL	OBL
JUNCUS BALTICUS WILLD.	FACW+	LIMOSELLA SUBULATA E. IVES	OBL
JUNCUS BREVICAUDATUS (ENGELM.) FERNALD	OBL	LINDERA BENZOIN (L.) BLUME	FACW
JUNCUS BUFONIUS L.	FACW	LINDERNIA ANAGALLIDEA (MICHX.) PENNELL	OBL
JUNCUS CANADENSIS J. GAY	OBL	LINDERNIA DUBIA (L.) PENNELL	OBL
JUNCUS DEBILIS GRAY	FACW	LINNAEA BOREALIS L.	FAC
JUNCUS DICHOTOMUS ELLIOTT	FACW+	LINUM MEDIUM (PLANCH.) BRITTON	FACU
JUNCUS EFFUSUS L.	FACW+	LINUM STRIATUM WALTER	FACW
JUNCUS GERARDII LOISELEUR	FAC	LINUM VIRGINIANUM L.	FACU
JUNCUS GREENEI OAKES & TUCKERMAN	FACW	LIPARIS LILIIFOLIA (L.) L.C. RICH. EX KER-GAWL	FACU-
JUNCUS MARGINATUS ROSTK.	OBL	LIPARIS LOESELII (L.) L.C. RICH.	FACW
JUNCUS MILITARIS BIGEL	OBL	LIRIODENDRON TULIPIFERA L.	FACU
JUNCUS NODOSUS L.	OBL	LISTERA CORDATA (L.) R. BR.	FACW+
JUNCUS PELOCARPUS E. MEYER	FAC	LOBELIA CARDINALIS L.	FACW+
JUNCUS PLATYPHYLLUS (WIEGAND) FERNALD	FACW	LOBELIA DORTMANNA L.	OBL
JUNCUS SCIRPOIDES LAM.	FACW	LOBELIA INFLATA L.	FACU
JUNCUS SECUNDUS BEAUV.	FACU-	LOBELIA SPICATA LAM.	FAC-
JUNCUS TENUIS WILLD.	FAC-	LOLIUM PERENNE L.	FACU-
JUNIPERUS HORIZONTALIS MOENCH	FACU	LONICERA DIOICA L.	FACU-
JUNIPERUS VIRGINIANA L.	FACU	LONICERA HIRSUTA A. EAT.	FACU
KALMIA ANGUSTIFOLIA L.	FAC	LONICERA JAPONICA THUNB.	FAC
KALMIA LATIFOLIA L.	FACU	LONICERA OBLONGIFOLIA (GOLDIE) HOOK.	FAC-
KALMIA POLIFOLIA WANGENH.	OBL	LONICERA SEMPERVIRENS L.	OBL
KICKXIA ELATINE (L.) DUMORT.	FAC	LOTUS CORNICULATUS L.	FACU-
LACHNANTHES CAROLINIANA (LAM.) DANDY	OBL	LUDWIGIA ALTERNIFOLIA L.	FACW+
LACTUCA BIENNIS (MOENCH) FERNALD	FACU-	LUDWIGIA PALUSTRIS (L.) ELLIOTT	OBL
LACTUCA CANADENSIS L.	FACU-		

Symbology: OBL (Obligate), FACW (Facultative Wetland), FAC (Facultative), FACU (Facultative Upland), NI (no indicator assigned), * (limited ecological information), + (higher portion of frequency range), and - (lower portion of frequency range). See discussion of hydrophyte definition and concept in Chapter 6.

Plant Species that Occur in Rhode Island's Wetlands (Continued)

GENUS-SPECIES-AUTHOR-TRINOMIAL-TRINOMIAL AUTHOR	RIIND	GENUS-SPECIES-AUTHOR-TRINOMIAL-TRINOMIAL AUTHOR	RIIND
LUDWIGIA SPHAEROCARPA ELLIOTT	OBL	MITCHELLA REPENS L.	FACU
LUDWIGIA X LACUSTRIS E. EAMES	OBL	MOEHRINGIA LATERIFLORA (L.) FENZL	FAC
LUZULA ACUMINATA RAF.	FAC	MOLLUGO VERTICILLATA L.	FAC
LUZULA MULTIFLORA (EHRH. EX HOFFM.) LEJ.	FACU	MONARDA DIDYMA L.	FAC+
LYCHNIS FLOS-CUCULI L.	FACU	MONOTROPA UNIFLORA L.	FACU-
LYCOPODIUM ANNOTINUM L.	FAC	MUHLENBERGIA FRONDOSA (POIR.) FERNALD	FAC
LYCOPODIUM APPRESSUM (CHAPM.) LLOYD & UNDERW.	FACW+	MUHLENBERGIA GLOMERATA (WILLD.) TRIN.	FACW
LYCOPODIUM CLAVATUM L.	FAC	MUHLENBERGIA MEXICANA (L.) TRIN.	FACW
LYCOPODIUM COMPLANATUM L.	FACU-	MUHLENBERGIA SCHREBERI J.F. GMEL.	FAC
LYCOPODIUM DENDROIDEUM MICHX.	FACU	MUHLENBERGIA SYLVATICA TORR. EX GRAY	FAC+
LYCOPODIUM INUNDATUM L.	OBL	MUHLENBERGIA UNIFLORA (MUHL.) FERNALD	OBL
LYCOPODIUM LUCIDULUM MICHX.	FACW	MYOSOTIS LAXA LEHM.	OBL
LYCOPODIUM OBSCURUM L.	FACU	MYOSOTIS SCORPIOIDES L.	OBL
LYCOPUS AMERICANUS MUHL. EX W. BARTON	OBL	MYOSOTIS VERNA NUTT.	FAC-
LYCOPUS AMPLECTENS RAF.	OBL	MYRICA GALE L.	OBL
LYCOPUS RUBELLUS MOENCH	OBL	MYRICA PENSYLVANICA LOISELEUR	FAC
LYCOPUS UNIFLORUS MICHX.	OBL	MYRIOPHYLLUM HETEROPHYLLUM MICHX.	OBL
LYCOPUS VIRGINICUS L.	OBL	MYRIOPHYLLUM HUMILE (RAF.) MORONG	OBL
LYCOPODIUM PALMATUM (BERNH.) SWARTZ	FACW	MYRIOPHYLLUM PINNATUM (WALTER) B.S.P.	OBL
LYONIA LIGUSTRINA (L.) DC.	FACW	MYRIOPHYLLUM TENELLUM BIGEL.	OBL
LYONIA MARIANA (L.) D. DON	FAC-	NAJAS FLEXILIS (WILLD.) ROSTK. & W.L.E. SCHMIDT	OBL
LYSIMACHIA CILIATA L.	FACW	NAJAS GRACILLIMA (A. BRAUN) MAGNUS	OBL
LYSIMACHIA HYBRIDA MICHX.	OBL	NAJAS GUADALUPENSIS (SPRENG.) MORONG	OBL
LYSIMACHIA NUMMULARIA L.	FACW	NASTURTIUM OFFICINALE R. BR. IN W.T. AIT.	OBL
LYSIMACHIA QUADRIFOLIA L.	FACU-	NEMOPANTHUS MUCRONATUS (L.) TRELEASE	OBL
LYSIMACHIA TERRESTRIS (L.) B.S.P.	OBL	NEPETA CATARIA L.	FACU
LYSIMACHIA THYRSIFLORA L.	OBL	NUPHAR LUTEUM (L.) SIBTH. & J.E. SMITH	OBL
LYSIMACHIA VULGARIS L.	FAC+	NYMPHAEA ODORATA SOLAND. IN AIT.	OBL
LYSIMACHIA X PRODUCTA (GRAY) FERNALD	FACW+	NYMPHOIDES CORDATA (ELLIOTT) FERNALD	OBL
LYTHRUM ALATUM PURSH	FACW+	NYSSA SYLVATICA MARSHALL	FAC
LYTHRUM HYSSOPIFOLIA L.	OBL	NYSSA SYLVATICA MARSHALL VAR. BIFLORA (WALTER) SARG.	FACW+
LYTHRUM SALICARIA L.	FACW+	OENOTHERA BIENNIS L.	FACU-
MAIANTHEMUM CANADENSE DESF.	FAC-	OENOTHERA FRUTICOSA L.	FAC
MALAXIS UNIFOLIA MICHX.	FAC	OENOTHERA PARVIFLORA L.	FACU-
MATRICARIA MATRICARIOIDES (LESS.) T. PORTER	FACU	OENOTHERA PERENNIS L.	FAC-
MATTEUCCIA STRUTHIOPTERIS (L.) TODARO	FACW	OMOCLEA SENSIBILIS L.	FACW
MEGALODONTA BECKII (TORR.) GREENE	OBL	OPHIOGLOSSUM VULGATUM L.	FACW
MELAMPYRUM LINEARE DESR.	FACU	ORNITHOGALUM UMBELLATUM L.	FACU
MELANTHIUM VIRGINICUM L.	FACW+	OROBANCHE UNIFLORA L.	FACU
MELILOTUS ALBA MEDIC.	FACU-	ORONTIUM AQUATICUM L.	OBL
MELILOTUS OFFICINALIS LAM.	FACU-	OSMORHIZA CLAYTONII (MICHX.) C.B. CLARKE	FACU-
MENTHA ARVENSIS L.	FACW	OSMORHIZA LONGISTYLIS (TORR.) DC.	FACU
MENTHA CARDIACA BAKER	FACW	OSMUNDA CINNAMOMEA L.	FACW
MENTHA CITRATA EHRH.	FACW+	OSMUNDA CLAYTONIANA L.	FAC
MENTHA SPICATA L.	FACW+	OSMUNDA REGALIS L.	OBL
MENTHA X PIPERITA L.	FACW+	OSTRYA VIRGINIANA (MILL.) K. KOCH	FACU-
MENYANTHES TRIFOLIATA L.	OBL	OXALIS CORNICULATA L.	FACU
MIKANIA SCANDENS (L.) WILLD.	FACW+	OXALIS MONTANA RAF.	FAC-
MIMULUS ALATUS AIT.	OBL	PANICUM AMARUM ELLIOTT	FACU-
MIMULUS RINGENS L.	OBL	PANICUM CAPILLARE L.	FAC-
MIRABILIS NYCTAGINEA (MICHX.) MACMIL.	FACU-	PANICUM DICHOTOMIFLORUM MICHX.	FACW-
MISCANTHUS SINENSIS ANDERSS.	FACU	PANICUM LONGIFOLIUM TORR.	OBL

Symbology: OBL (Obligate), FACW (Facultative Wetland), FAC (Facultative), FACU (Facultative Upland), NI (no indicator assigned), * (limited ecological information), + (higher portion of frequency range), and − (lower portion of frequency range). See discussion of hydrophyte definition and concept in Chapter 6.

Plant Species that Occur in Rhode Island's Wetlands (*Continued*)

GENUS-SPECIES-AUTHOR-TRINOMIAL-TRINOMIAL AUTHOR	RIIND	GENUS-SPECIES-AUTHOR-TRINOMIAL-TRINOMIAL AUTHOR	RIIND
PANICUM RIGIDULUM BOSC EX NEES	FACW+	POA PRATENSIS L.	FACU
PANICUM TUCKERMANII FERNALD	FAC-	POA TRIVIALIS L.	FACU
PANICUM VERRUCOSUM MUHL.	FACW	PODOPHYLLUM PELTATUM L.	FACU
PANICUM VIRGATUM L.	FAC	PODOSTEMUM CERATOPHYLLUM MICHX.	OBL
PARIETARIA PENSYLVANICA MUHL. EX WILLD.	FACU-	POGONIA OPHIOGLOSSOIDES (L.) JUSS.	OBL
PARNASSIA GLAUCA RAF.	OBL	POLEMONIUM REPTANS L.	FACU
PARTHENOCISSUS QUINQUEFOLIA (L.) PLANCH.	FACU	POLYGALA BREVIFOLIA NUTT.	OBL
PARTHENOCISSUS VITACEA (KNERR) A. HITCHC.	FACU	POLYGALA CRUCIATA L.	FACW+
PASPALUM LAEVE MICHX.	FAC+	POLYGALA NUTTALLII TORR. & GRAY	FAC
PASPALUM SETACEUM MICHX.	FACU+	POLYGALA PAUCIFOLIA WILLD.	FACU
PEDICULARIS CANADENSIS L.	FACU	POLYGALA SANGUINEA L.	FACU
PEDICULARIS LANCEOLATA MICHX.	FACW	POLYGALA SENEGA L.	FACU
PELTANDRA VIRGINICA (L.) KUNTH	OBL	POLYGONATUM BIFLORUM (WALTER) ELLIOTT	FACU
PENSTEMON DIGITALIS NUTT.	FAC	POLYGONATUM COMMUTATUM (J.A. & J.H. SCHULTES) A. DIETR.	FACU
PENSTEMON LAEVIGATUS SOLAND.	FACU	POLYGONUM AMPHIBIUM L.	OBL
PENTHORUM SEDOIDES L.	OBL	POLYGONUM ARIFOLIUM L.	OBL
PHALARIS ARUNDINACEA L.	FACW+	POLYGONUM AVICULARE L.	FACU
PHALARIS CANARIENSIS L.	FACU	POLYGONUM CAREYI OLNEY	FACW
PHLEUM PRATENSE L.	FACU	POLYGONUM CONVOLVULUS L.	FACU
PHLOX PANICULATA L.	FACW	POLYGONUM CUSPIDATUM SIEBOLD & ZUCCAR.	FACU-
PHRAGMITES AUSTRALIS (CAV.) TRIN. EX STEUD.	FAC	POLYGONUM ERECTUM L.	FACU
PHYSALIS ANGULATA L.	FACU-	POLYGONUM GLAUCUM NUTT.	OBL
PHYSALIS PUBESCENS L.	FACU-	POLYGONUM HYDROPIPER L.	OBL
PHYSOCARPUS OPULIFOLIUS (L.) MAXIM.	FAC+	POLYGONUM HYDROPIPEROIDES MICHX.	OBL
PHYSOSTEGIA VIRGINIANA (L.) BENTH.	FACU+	POLYGONUM LAPATHIFOLIUM L.	FACW+
PHYTOLACCA AMERICANA L.	FACU	POLYGONUM OPELOUSANUM RIDDELL EX SMALL	OBL
PICEA GLAUCA (MOENCH) VOSS	FACU	POLYGONUM ORIENTALE L.	FACU-
PICEA MARIANA (MILL.) B.S.P.	FACW	POLYGONUM PENSYLVANICUM L.	FACW
PILEA PUMILA (L.) GRAY	FACW	POLYGONUM PERSICARIA L.	FACW
PINUS RESINOSA SOLAND. IN AIT.	FAC	POLYGONUM PUNCTATUM ELLIOTT	OBL
PINUS RIGIDA MILL.	FAC	POLYGONUM RAMOSISSIMUM MICHX.	FAC
PINUS STROBUS L.	FACU	POLYGONUM ROBUSTIUS (SMALL) FERNALD	OBL
PLANTAGO MAJOR L.	FACU	POLYGONUM SAGITTATUM L.	OBL
PLANTAGO MARITIMA L.	FACW	POLYGONUM SCANDENS L.	FAC
PLANTAGO RUGELII DECNE.	FACU	POLYGONUM SETACEUM BALDW.	OBL
PLATANTHERA BLEPHARIGLOTTIS (WILLD.) LINDL.	OBL	POLYGONUM VIRGINIANUM L.	FAC
PLATANTHERA CILIARIS (L.) LINDL.	FACW	POLYSTICHUM ACROSTICHOIDES (MICHX.) SCHOTT	FACU-
PLATANTHERA FLAVA (L.) LINDL.	FACW	PONTEDERIA CORDATA L.	OBL
PLATANTHERA GRANDIFLORA (BIGEL.) LINDL.	FACW	POPULUS BALSAMIFERA L.	FACW
PLATANTHERA HOOKERI (TORR.) LINDL.	FAC	POPULUS DELTOIDES N. BARTRAM EX MARSHALL	FACW
PLATANTHERA HYPERBOREA (L.) LINDL.	FACW	POPULUS GRANDIDENTATA MICHX.	FAC
PLATANTHERA LACERA (MICHX.) G. DON	FACW	POPULUS TREMULA L.	FACU-
PLATANTHERA ORBICULATA (PURSH) LINDL.	FAC	PORTULACA OLERACEA L.	FACU
PLATANTHERA PSYCHODES (L.) LINDL.	FACW+	POTAMOGETON AMPLIFOLIUS TUCKERMAN	FAC
PLATANTHERA X CLAVELLATA (MICHX.) LUER	FACW-	POTAMOGETON BICUPULATUS FERNALD	OBL
PLATANUS OCCIDENTALIS L.	OBL	POTAMOGETON CONFERVOIDES REICHENB.	OBL
PLUCHEA PURPURASCENS (SWARTZ) DC.	FACU-	POTAMOGETON CRISPUS L.	OBL
POA ANGUSTIFOLIA L.	FACU	POTAMOGETON EPIHYDRUS RAF.	OBL
POA ANNUA L.	FACU	POTAMOGETON FOLIOSUS RAF.	2V OBL
POA COMPRESSA L.	FAC	POTAMOGETON GRAMINEUS L.	OBL
POA NEMORALIS L.	FACU	POTAMOGETON NATANS L.	OBL
POA PALUSTRIS L.	FACW	POTAMOGETON NODOSUS POIR.	OBL

A-8

Symbology: OBL (Obligate), FACW (Facultative Wetland), FAC (Facultative), FACU (Facultative Upland), NI (no indicator assigned), * (limited ecological information), + (higher portion of frequency range), and − (lower portion of frequency range). See discussion of hydrophyte definition and concept in Chapter 6.

Plant Species that Occur in Rhode Island's Wetlands (Continued)

GENUS-SPECIES-AUTHOR-TRINOMIAL-TRINOMIAL AUTHOR	R1IND
POTAMOGETON OAKESIANUS J.W. ROBBINS	OBL
POTAMOGETON OBTUSIFOLIUS F. MERTENS & W. KOCH	OBL
POTAMOGETON PECTINATUS L	OBL
POTAMOGETON PERFOLIATUS L	OBL
POTAMOGETON PULCHER TUCKERMAN	OBL
POTAMOGETON PUSILLUS L	OBL
POTAMOGETON ROBBINSII OAKES	OBL
POTAMOGETON SPIRILLUS TUCKERMAN	OBL
POTAMOGETON VASEYI J.W. ROBBINS	OBL
POTAMOGETON ZOSTERIFORMIS FERNALD	OBL
POTENTILLA ANSERINA L	OBL
POTENTILLA FRUTICOSA L	FACW
POTENTILLA NORVEGICA L	FACU
POTENTILLA SIMPLEX MICHX.	FACU-
PRENANTHES ALBA L	FACU
PRENANTHES ALTISSIMA L	FACU-
PROSERPINACA PALUSTRIS L	OBL
PROSERPINACA PECTINATA LAM.	OBL
PROSERPINACA X INTERMEDIA MACKENZ.	OBL
PRUNELLA VULGARIS L	FACU+
PRUNUS AMERICANA MARSHALL	FACU-
PRUNUS PENSYLVANICA L.F.	FACU-
PRUNUS SEROTINA EHRH.	FACU
PRUNUS VIRGINIANA L	FACU
PSILOCARYA SCIRPOIDES TORR.	OBL
PTELEA TRIFOLIATA L	FAC
PTERIDIUM AQUILINUM (L) KUHN	FACU
PTILIMNIUM CAPILLACEUM (MICHX.) RAF.	OBL
PUCCINELLIA DISTANS (L.) PARLAT.	OBL
PUCCINELLIA FASCICULATA (TORR.) BICKN.	OBL
PUCCINELLIA LANGEANA (BERLIN) SORENS. EX HULTEN	FACW+
PUCCINELLIA MARITIMA (HUDS.) PARLAT.	OBL
PUCCINELLIA PALLIDA (TORR.) R.T. CLAUSEN	OBL
PUCCINELLIA PUMILA (VASEY) A. HITCHC.	FACW
PYCNANTHEMUM MUTICUM (MICHX.) PERS.	FACW
PYCNANTHEMUM TENUIFOLIUM SCHRAD.	FACW
PYCNANTHEMUM VERTICILLATUM (MICHX.) PERS.	FAC
PYCNANTHEMUM VIRGINIANUM (L.) TH. DURAND&B. D. JACKS. EX B. ROB. &FERNALD	FAC
PYROLA ROTUNDIFOLIA L	FAC
PYROLA SECUNDA L	FAC
PYROLA UNIFLORA L	FAC
QUERCUS ALBA L	FACU-
QUERCUS BICOLOR WILLD.	FACW+
QUERCUS FALCATA MICHX.	FACU-
QUERCUS MACROCARPA MICHX.	FAC-
QUERCUS PALUSTRIS MUENCHH.	FACW
QUERCUS PRINOIDES WILLD.	NI
QUERCUS RUBRA L	FACU-
RANUNCULUS ABORTIVUS L	FACW-
RANUNCULUS ACRIS L	FAC+
RANUNCULUS ALLEGHENIENSIS BRITTON	FAC
RANUNCULUS AMBIGENS S. WATS.	OBL

GENUS-SPECIES-AUTHOR-TRINOMIAL-TRINOMIAL AUTHOR	R1IND
RANUNCULUS AQUATILIS L	OBL
RANUNCULUS CYMBALARIA PURSH	OBL
RANUNCULUS FASCICULARIS MUHL EX BIGEL	FACU
RANUNCULUS FLABELLARIS RAF.	OBL
RANUNCULUS MICRANTHUS NUTT.	FACU
RANUNCULUS PENSYLVANICUS L.F.	OBL
RANUNCULUS RECURVATUS POIR.	FAC+
RANUNCULUS REPENS L	FAC
RANUNCULUS SCELERATUS L	OBL
RANUNCULUS TRICHOPHYLLUS D. CHAIX	OBL
RHEXIA VIRGINICA L	OBL
RHINANTHUS CRISTA-GALLI L	FAC
RHODODENDRON CANADENSE (L.) B.S.P.	FACW
RHODODENDRON MAXIMUM L	FAC
RHODODENDRON PERICLIMENOIDES (MICHX.) SHINNERS	FAC
RHODODENDRON PRINOPHYLLUM (SMALL) MILLAIS	FAC
RHODODENDRON VISCOSUM (L) TORR.	OBL
RHUS COPALLINUM L	NI
RHYNCHOSPORA ALBA (L.) VAHL	OBL
RHYNCHOSPORA CAPITELLATA (MICHX.) VAHL	OBL
RHYNCHOSPORA FUSCA (L.) W.T. AIT.	OBL
RHYNCHOSPORA INUNDATA (OAKES) FERNALD	OBL
RHYNCHOSPORA MACROSTACHYA TORR.	OBL
RHYNCHOSPORA TORREYANA GRAY	FACW+
RIBES AMERICANUM MILL	FACW
RIBES HIRTELLUM MICHX.	FAC
RIBES ODORATUM H.L. WENDL	FACU
RIBES TRISTE PALLAS	FACU-
ROBINIA PSEUDOACACIA L	OBL
RORIPPA PALUSTRIS (L.) BESSER	FACW
RORIPPA SYLVESTRIS (L.) BESSER	FAC
ROSA ACICULARIS LINDL U	FACU
ROSA BLANDA AIT.	FACU
ROSA MICRANTHA J.E. SMITH	FACU
ROSA MULTIFLORA THUNB.	FACU
ROSA NITIDA WILLD.	FACW+
ROSA PALUSTRIS MARSHALL	OBL
ROSA RUGOSA THUNB.	FACU-
ROSA VIRGINIANA MILL	FAC
ROTALA RAMOSIOR (L.) KOEHNE	OBL
RUBUS ALLEGHENIENSIS T. PORTER	FACU-
RUBUS ALUMNUS L.H. BAILEY	FACU-
RUBUS ASCENDENS N.H. BLANCH.	FAC
RUBUS ENSLENII TRATT.	FACU
RUBUS FLORICOMUS N.H. BLANCH.	FACU
RUBUS HISPIDUS L	FACW
RUBUS IDAEUS L	FAC-
RUBUS LAWRENCEI L.H. BAILEY	OBL
RUBUS PUBESCENS RAF.	FACW
RUBUS SEMISETOSUS N.H. BLANCH.	FAC
RUBUS SETOSUS BIGEL	FACW+
RUBUS STRIGOSUS MICHX.	NI

Symbology: OBL (Obligate), FACW (Facultative Wetland), FAC (Facultative), FACU (Facultative Upland), NI (no indicator assigned), * (limited ecological information), + (higher portion of frequency range), and − (lower portion of frequency range). See discussion of hydrophyte definition and concept in Chapter 6.

Plant Species that Occur in Rhode Island's Wetlands (Continued)

GENUS-SPECIES-AUTHOR-TRINOMIAL-TRINOMIAL AUTHOR	R1IND	GENUS-SPECIES-AUTHOR-TRINOMIAL-TRINOMIAL AUTHOR	R1IND
RUBUS X GROUTIANUS N.H. BLANCH.	FAC	SANICULA GREGARIA BICKN.	FACU
RUDBECKIA HIRTA L.	FACU-	SANICULA MARILANDICA L.	NI
RUMEX ALTISSIMUS A. WOOD	FACW-	SAPONARIA OFFICINALIS L.	FACU-
RUMEX CRISPUS L.	FACU	SARRACENIA PURPUREA L.	OBL
RUMEX DOMESTICUS HARTM.	FAC	SASSAFRAS ALBIDUM (NUTT.) NEES	FACU-
RUMEX MARITIMUS L.	FACW	SAURURUS CERNUUS L.	OBL
RUMEX OBTUSIFOLIUS L.	FACU-	SAXIFRAGA PENSYLVANICA L.	OBL
RUMEX OBICULATUS GRAY	OBL	SAXIFRAGA VIRGINIENSIS MICHX.	FAC-
RUMEX TRIANGULIVALVIS (DANSER) RECH. F.	FACU	SCHEUCHZERIA PALUSTRIS L.	OBL
RUMEX VERTICILLATUS L.	OBL	SCHIZACHNE PURPURASCENS (TORR.) SWALLEN	FACU-
RUPPIA MARITIMA L.	OBL	SCHIZACHYRIUM SCOPARIUM (MICHX.) NASH	FACU-
SABATIA DODECANDRA (L.) B.S.P.	OBL	SCIRPUS ACUTUS MUHL EX BIGEL	OBL
SABATIA KENNEDYANA FERNALD	OBL	SCIRPUS AMERICANUS PERS.	OBL
SABATIA STELLARIS PURSH	FACW+	SCIRPUS ATROCINCTUS FERNALD	FACW+
SAGINA DECUMBENS (ELLIOTT) TORR. & GRAY	FAC	SCIRPUS ATROVIRENS WILLD.	OBL
SAGINA PROCUMBENS L.	FACW-	SCIRPUS CYPERINUS (L.) KUNTH	FACW+
SAGITTARIA CALYCINA ENGELM.	OBL	SCIRPUS EXPANSUS FERNALD	OBL
SAGITTARIA ENGELMANNIANA J. G. SMITH	OBL	SCIRPUS FLUVIATILIS (TORR.) GRAY	OBL
SAGITTARIA GRAMINEA MICHX.	OBL	SCIRPUS GEORGIANUS R.M. HARPER	OBL
SAGITTARIA LATIFOLIA WILLD.	OBL	SCIRPUS HETEROCHAETUS CHASE	OBL
SAGITTARIA RIGIDA PURSH	OBL	SCIRPUS MARITIMUS L.	OBL
SAGITTARIA STAGNORUM SMALL	OBL	SCIRPUS MICROCARPUS J. & K. PRESL	OBL
SAGITTARIA SUBULATA (L.) BUCHENAU	OBL	SCIRPUS PECKII BRITTON	OBL
SALICORNIA BIGELOVII TORR.	OBL	SCIRPUS PEDICELLATUS FERNALD	OBL
SALICORNIA EUROPAEA L.	OBL	SCIRPUS POLYPHYLLUS VAHL	OBL
SALICORNIA VIRGINICA L.	FACW	SCIRPUS PURSHIANUS FERNALD	OBL
SALIX ALBA L.	FACW-	SCIRPUS ROBUSTUS PURSH	OBL
SALIX AMYGDALOIDES ANDERSS.	FACW	SCIRPUS SMITHII GRAY	OBL
SALIX BABYLONICA L.	FACW	SCIRPUS SUBTERMINALIS TORR.	OBL
SALIX BEBBIANA SARG.	FACW	SCIRPUS TORREYI OLNEY	OBL
SALIX DISCOLOR MUHL.	FACW	SCIRPUS VALIDUS VAHL	OBL
SALIX ERIOCEPHALA MICHX.	OBL	SCLERANTHUS ANNUUS L.	FACU
SALIX EXIGUA NUTT.	FAC+	SCLERIA RETICULARIS MICHX.	OBL
SALIX FRAGILIS L.	FACU	SCLERIA TRIGLOMERATA MICHX.	FAC
SALIX HUMILIS MARSHALL	FACW	SCLEROLEPIS UNIFLORA (WALTER) B.S.P.	OBL
SALIX LUCIDA MUHL.	FAC	SCROPHULARIA LANCEOLATA PURSH	FACU+
SALIX MYRICOIDES MUHL.	FACW+	SCROPHULARIA MARILANDICA L	FACU-
SALIX NIGRA MARSHALL	OBL	SCUTELLARIA GALERICULATA L	OBL
SALIX PEDICELLARIS PURSH	OBL	SCUTELLARIA INTEGRIFOLIA L	FACW
SALIX PETIOLARIS J.E. SMITH	OBL	SCUTELLARIA LATERIFLORA L.	FACW+
SALIX PURPUREA L.	NI	SELAGINELLA APODA (L.) SPRING	FACW
SALIX RIGIDA MUHL	OBL	SENECIO AUREUS L.	FACW
SALIX SERICEA MARSHALL	OBL	SENECIO OBOVATUS MUHL EX WILLD.	FACU-
SALIX VIMINALIS L.	FACW-	SENECIO PAUPERCULUS MICHX.	FAC
SALSOLA KALI L.	FACU-	SENECIO VULGARIS L.	FAC
SALSOLA PESTIFER A. NELS.	FACU-	SETARIA GENICULATA (LAM.) BEAUV.	FAC
SAMBUCUS CANADENSIS L	FACW-	SETARIA GLAUCA (L.) BEAUV.	FACU
SAMBUCUS RACEMOSA L.	FACU	SETARIA ITALICA (L.) BEAUV.	FACU
SAMOLUS PARVIFLORUS RAF.	OBL	SETARIA VERTICILLATA (L.) BEAUV.	FAC
SANGUINARIA CANADENSIS L.	NI	SICYOS ANGULATUS L.	FACU
SANGUISORBA CANADENSIS L	FACW+	SISYMBRIUM ALTISSIMUM L.	FACU-
SANGUISORBA MINOR SCOP.	FAC	SISYRINCHIUM ANGUSTIFOLIUM MILL	FACW-

Symbology: OBL (Obligate), FACW (Facultative Wetland), FAC (Facultative), FACU (Facultative Upland), NI (no indicator assigned), * (limited ecological information), + (higher portion of frequency range), and − (lower portion of frequency range). See discussion of hydrophyte definition and concept in Chapter 6.

Plant Species that Occur in Rhode Island's Wetlands (Continued)

GENUS-SPECIES-AUTHOR-TRINOMIAL AUTHOR	R1IND	GENUS-SPECIES-AUTHOR-TRINOMIAL-TRINOMIAL AUTHOR	R1IND
SISYRINCHIUM ARENICOLA BICKN.	FACU	SPIRANTHES ODORATA (NUTT.) LINDL.	OBL
SISYRINCHIUM ATLANTICUM BICKN.	FACW	SPIRANTHES PRAECOX (WALTER) S. WATS.	OBL
SISYRINCHIUM MONTANUM GREENE	FAC	SPIRANTHES ROMANZOFFIANA CHAM.	OBL
SISYRINCHIUM MUCRONATUM MICHX.	FAC+	SPIRANTHES VERNALIS ENGELM. & GRAY	FAC
SIUM CARSONII E.M. DURAND EX GRAY	OBL	SPIRODELA POLYRHIZA (L.) SCHLEID.	OBL
SIUM SUAVE WALTER	OBL	STACHYS ASPERA MICHX.	FAC
SMILACINA RACEMOSA (L.) DESF.	FACU-	STACHYS CORDATA RIDDELL	FAC
SMILACINA STELLATA (L.) DESF.	FACW	STACHYS HISPIDA PURSH	OBL
SMILACINA TRIFOLIA (L.) DESF.	OBL	STACHYS HYSSOPIFOLIA MICHX.	FACW+
SMILAX GLAUCA WALTER	FACU	STACHYS PALUSTRIS L.	OBL
SMILAX HERBACEA L.	FAC	STACHYS TENUIFOLIA WILLD.	FACW+
SMILAX PULVERULENTA MICHX.	FACU	STAPHYLEA TRIFOLIA L.	FAC
SMILAX ROTUNDIFOLIA L.	FAC	STELLARIA CALYCANTHA (LEDEB.) BONG.	FACW
SOLANUM AMERICANUM MILL.	FACU-	STELLARIA GRAMINEA L.	FACU-
SOLANUM DULCAMARA L.	FAC-	STELLARIA LONGIFOLIA MUHL EX WILLD.	FACW
SOLANUM NIGRUM L.	FACU-	STREPTOPUS ROSEUS MICHX.	FACU-
SOLIDAGO ALTISSIMA L.	FACU-	STROPHOSTYLES HELVOLA (L.) ELLIOTT	FACU
SOLIDAGO CAESIA L.	FACU	SUAEDA AMERICANA (PERS.) FERNALD	OBL
SOLIDAGO CANADENSIS L.	FACU	SUAEDA LINEARIS (ELLIOTT) MOQ.	OBL
SOLIDAGO ELLIOTTII TORR. & GRAY	OBL	SUAEDA MARITIMA (L.) DUMORT.	OBL
SOLIDAGO FLEXICAULIS L.	FACU	SYMPHORICARPOS ALBUS (L.) BLAKE	FACU-
SOLIDAGO GIGANTEA AIT.	FACW	SYMPLOCARPUS FOETIDUS (L.) SALISB.	OBL
SOLIDAGO NUTTALLII GREENE	FACU+	TARAXACUM OFFICINALE G.H. WEBER	FACU-
SOLIDAGO PUBERULA NUTT.	FACU-	TAXUS CANADENSIS MARSHALL	FACU
SOLIDAGO RUGOSA MILL.	FAC	TEUCRIUM CANADENSE L.	FACW
SOLIDAGO SEMPERVIRENS L.	FACW	THALICTRUM DASYCARPUM FISCH. & AVE-LALL	FAC
SOLIDAGO SPATHULATA DC.	FACU	THALICTRUM DIOICUM L.	FACU-
SOLIDAGO STRICTA AIT.	FACW	THALICTRUM PUBESCENS PURSH	FACW+
SOLIDAGO ULIGINOSA NUTT.	OBL	THELYPTERIS HEXAGONOPTERA (MICHX.) WEATHERBY	FAC
SOLIDAGO X ASPERULA DESF.	OBL*	THELYPTERIS NOVEBORACENSIS (L.) NIEUWL.	FACU
SONCHUS ASPER (L.) J. HILL	FAC	THELYPTERIS SIMULATA (DAVENP.) NIEUWL.	FACW
SPARGANIUM AMERICANUM NUTT.	OBL	THELYPTERIS THELYPTEROIDES (MICHX.) J. HOLUB	FACW+
SPARGANIUM ANDROCLADUM (ENGELM.) MORONG	OBL	THLASPI ARVENSE L.	NI
SPARGANIUM CHLOROCARPUM RYDB.	OBL	THUJA OCCIDENTALIS L.	FACW
SPARGANIUM EURYCARPUM ENGELM. EX GRAY	OBL	TILIA AMERICANA L.	FACU
SPARGANIUM MINIMUM (HARTM.) FR.	OBL	TOXICODENDRON QUERCIFOLIA (MICHX.) KUNTZE	FAC
SPARTINA ALTERNIFLORA LOISELEUR	OBL	TOXICODENDRON RADICANS (L.) KUNTZE	FAC
SPARTINA CAESPITOSA A.A. EAT.	FACW+	TOXICODENDRON RYDBERGII (SMALL EX RYDB.) GREENE	FAC-
SPARTINA PATENS (AIT.) MUHL.	OBL	TOXICODENDRON VERNIX (L.) KUNTZE	OBL
SPARTINA PECTINATA LINK	OBL	TRIADENUM FRASERI (SPACH) GLEASON	OBL
SPERGULARIA CANADENSIS (PERS.) D. DON	OBL	TRIADENUM VIRGINICUM (L.) RAF.	OBL
SPERGULARIA MARINA (L.) GRISEB.	OBL	TRIDENS FLAVUS (L.) A. HITCHC.	FACU*
SPERGULARIA RUBRA (L.) J. & K. PRESL	FACU	TRIENTALIS BOREALIS RAF.	FAC
SPHENOPHOLIS OBTUSATA (MICHX.) SCRIBN.	FAC-	TRIFOLIUM HYBRIDUM L.	FACU-
SPHENOPHOLIS PENSYLVANICA (L.) A. HITCHC.	OBL	TRIFOLIUM PRATENSE L.	FACU-
SPIRAEA JAPONICA L.F.	FACU	TRIFOLIUM REPENS L.	FACU-
SPIRAEA LATIFOLIA (AIT.) BORKH.	FAC+	TRIGLOCHIN MARITIMUM L.	OBL
SPIRAEA TOMENTOSA L.	FACW	TRIGLOCHIN PALUSTRE L.	OBL
SPIRANTHES CERNUA (L.) L.C. RICH.	FACW	TRILLIUM CERNUUM L.	FACW
SPIRANTHES GRAYI AMES	FACU-	TRILLIUM ERECTUM L.	FACU-
SPIRANTHES LACERA (RAF.) RAF.	FACU-	TRILLIUM UNDULATUM WILLD.	FACU*
SPIRANTHES LUCIDA (H.H. EAT.) AMES	FACW	TRIODANIS PERFOLIATA (L.) NIEUWL.	FAC

Symbology: OBL (Obligate), FACW (Facultative Wetland), FAC (Facultative), FACU (Facultative Upland), NI (no indicator assigned), * (limited ecological information), + (higher portion of frequency range), and − (lower portion of frequency range). See discussion of hydrophyte definition and concept in Chapter 6.

Plant Species that Occur in Rhode Island's Wetlands (Continued)

GENUS-SPECIES-AUTHOR-TRINOMIAL AUTHOR	R IND	GENUS-SPECIES-AUTHOR-TRINOMIAL AUTHOR	R IND
TRIPSACUM DACTYLOIDES (L.) L.	FACW	VIOLA BLANDA WILLD.	FACW
TSUGA CANADENSIS (L.) CARRIERE	FACU	VIOLA CONSPERSA REICHENB.	FACW
TUSSILAGO FARFARA L.	FACU	VIOLA CUCULLATA AIT.	FACW+
TYPHA ANGUSTIFOLIA L.	OBL	VIOLA INCOGNITA BRAINERD	FACW
TYPHA LATIFOLIA L.	OBL	VIOLA LANCEOLATA L.	OBL
TYPHA X GLAUCA GODR.	OBL	VIOLA NEPHROPHYLLA GREENE	FACW
ULMUS AMERICANA L.	FACW-	VIOLA PALLENS (BANKS) BRAINERD	OBL
ULMUS RUBRA MUHL.	FAC	VIOLA PAPILIONACEA PURSH	FAC
URTICA DIOICA L.	FACU	VIOLA PENSYLVANICA MICHX.	FACU
UTRICULARIA BIFLORA LAM.	OBL	VIOLA PRIMULIFOLIA L.	FAC+
UTRICULARIA CORNUTA MICHX.	OBL	VIOLA PUBESCENS AIT.	FACU-
UTRICULARIA GEMINISCAPA L. BENJ.	OBL	VIOLA ROTUNDIFOLIA MICHX.	FAC+
UTRICULARIA GIBBA L.	OBL	VIOLA SAGITTATA AIT.	FACW
UTRICULARIA INTERMEDIA HAYNE	OBL	VIOLA SEPTENTRIONALIS GREENE	FACU
UTRICULARIA MACRORHIZA LECONTE	OBL	VIOLA SORORIA WILLD.	FAC-
UTRICULARIA MINOR L.	OBL	VIOLA STRIATA AIT.	FACW
UTRICULARIA PURPUREA WALTER	OBL	VITIS AESTIVALIS MICHX.	FACU
UTRICULARIA RADIATA SMALL	OBL	VITIS LABRUSCA L.H. BAILEY	FACU
UTRICULARIA RESUPINATA B. GREENE	OBL	VITIS NOVAE-ANGLIAE FERNALD	NI
UTRICULARIA SUBULATA L.	FACU	VITIS RIPARIA MICHX.	FACW
UVULARIA PERFOLIATA L.	FACU-	VITIS VULPINA L.	FAC
UVULARIA SESSILIFOLIA L.	FACU-	WOODWARDIA AREOLATA (L.) T. MOORE	FACW+
VACCINIUM ANGUSTIFOLIUM AIT.	OBL	WOODWARDIA VIRGINICA (L.) J.E. SMITH	OBL
VACCINIUM CAESARIENSE MACKENZ.	FACW	XANTHIUM SPINOSUM L.	FACU
VACCINIUM CORYMBOSUM L.	OBL	XANTHIUM STRUMARIUM L.	FAC
VACCINIUM MACROCARPON AIT.	OBL	XYRIS CAROLINIANA WALTER	FACW+
VACCINIUM OXYCOCCOS L.	FACU-	XYRIS DIFFORMIS CHAPM.	OBL
VACCINIUM STAMINEUM L.	OBL	XYRIS MONTANA RIES	OBL
VALLISNERIA AMERICANA MICHX.	FACW+	XYRIS SMALLIANA NASH	OBL
VERATRUM VIRIDE AIT.	FACW+	XYRIS TORTA J.E. SMITH	OBL
VERBENA HASTATA L.	FACW+	ZANNICHELLIA PALUSTRIS L.	OBL
VERBENA OFFICINALIS L.	FACU	ZIZANIA AQUATICA L.	OBL
VERBENA URTICIFOLIA L.	FAC	ZIZIA APTERA (GRAY) FERNALD	FAC
VERBESINA ALTERNIFOLIA (L.) BRITTON	FACU-	ZIZIA AUREA (L.) W. KOCH	FAC
VERBESINA ENCELIOIDES (CAV.) BENTH. & HOOK. EX GRAY	FACW+	ZOSTERA MARINA L.	OBL
VERONIA NOVEBORACENSIS (L.) MICHX.	OBL		
VERONICA AMERICANA SCHWEINITZ EX BENTH.	OBL		
VERONICA ANAGALLIS-AQUATICA L.	NI		
VERONICA ARVENSIS L.	FACU-		
VERONICA OFFICINALIS L.	FACU-		
VERONICA PEREGRINA L.	OBL		
VERONICA SCUTELLATA L.	FACW		
VERONICA SERPYLLIFOLIA L.	FAC+		
VERONICASTRUM VIRGINICUM (L.) FARW.	FACU		
VIBURNUM CASSINOIDES L.	FACW		
VIBURNUM DENTATUM L.	FAC		
VIBURNUM LANTANOIDES MICHX.	FAC		
VIBURNUM LENTAGO L.	FAC		
VIBURNUM NUDUM L.	OBL		
VIBURNUM RECOGNITUM FERNALD	FACW-		
VICIA SATIVA L.	FACU-		
VIOLA ADUNCA J.E. SMITH	FAC		

Symbology: OBL (Obligate), FACW (Facultative Wetland), FAC (Facultative), FACU (Facultative Upland), NI (no indicator assigned), * (limited ecological information), + (higher portion of frequency range), and - (lower portion of frequency range). See discussion of hydrophyte definition and concept in Chapter 6.

As the Nation's principal conservation agency, the Department of the Interior has responsibility for most of our nationally owned public lands and natural resources. This includes fostering the wisest use of our land and water resources, protecting our fish and wildlife, preserving the environmental and cultural values of our national parks and historical places, and providing for the enjoyment of life through outdoor recreation. The Department assesses our energy and mineral resources and works to assure that their development is in the best interests of all our people. The Department also has a major responsibility for American Indian reservation communities and for people who live in island territories under U.S. administration.

U.S. Department of the Interior
Fish and Wildlife Service
One Gateway Center
Newton Corner, MA 02158

THIRD CLASS BOOK RATE